CATEGORICAL DATA ANALYSIS BY EXAMPLE

CATEGORICAL DATA ANALYSIS BY EXAMPLE

GRAHAM J. G. UPTON

Library of Congress Cataloging-in-Publication Data

Names: Upton, Graham J. G., author.
Title: Categorical data analysis by example / Graham J.G. Upton.
Description: Hoboken, New Jersey : John Wiley & Sons, 2016. | Includes index.
Identifiers: LCCN 2016031847 (print) | LCCN 2016045176 (ebook) | ISBN 9781119307860 (cloth) |
    ISBN 9781119307914 (pdf) | ISBN 9781119307938 (epub)
Subjects: LCSH: Multivariate analysis. | Log-linear models.
Classification: LCC QA278 .U68 2016 (print) | LCC QA278 (ebook) | DDC 519.5/35–dc23
LC record available at https://lccn.loc.gov/2016031847

Printed in the United States of America

SKY10092999_121124

CONTENTS

PREFACE

This book is aimed at all those who wish to discover how to analyze categorical data without getting immersed in complicated mathematics and without needing to wade through a large amount of prose. It is aimed at researchers with their own data ready to be analyzed and at students who would like an approachable alternative view of the subject. The few starred sections provide background details for interested readers, but can be omitted by readers who are more concerned with the "How" than the "Why."

As the title suggests, each new topic is illustrated with an example. Since the examples were as new to the writer as they will be to the reader, in many cases I have suggested preliminary visualizations of the data or informal analyses prior to the formal analysis. Any model provides, at best, a convenient simplification of a mass of data into a few summary figures. For a proper analysis of any set of data, it is essential to understand the background to the data and to have available information on all the relevant variables. Examples in textbooks cannot be expected to provide detailed insights into the data analyzed: those insights should be provided by the users of the book in the context of their own sets of data.

In many cases (particularly in the later chapters), R code is given and excerpts from the resulting output are presented. R was chosen simply because it is free! The thrust of the book is about the methods of analysis, rather than any particular programming language. Users of other languages (SAS, STATA, ...) would obtain equivalent output from their analyses; it would simply be presented in a slightly different format. The author does not claim to be an expert R programmer, so the example code can doubtless be improved. However, it should work adequately as it stands.

In the context of log-linear models for cross-tabulations, two "specialties of the house" have been included: the use of cobweb diagrams to get visual information concerning significant interactions, and a procedure for detecting outlier category combinations. The R code used for these is available and may be freely adapted.

GRAHAM J. G. UPTON

Wivenhoe, Essex
March, 2016

ACKNOWLEDGMENTS

A first thanks go to generations of students who have sat through lectures related to this material without complaining too loudly!

I have gleaned data from a variety of sources and particular thanks are due to Mieke van Hemelrijck and Sabine Rohrmann for making the NHANES III data available. The data on the hands of blues guitarists have been taken from the *Journal of Statistical Education*, which has an excellent online data resource. Most European and British data were abstracted from the UK Data Archive, which is situated at the University of Essex; I am grateful for their assistance and their permission to use the data. Those interested in election data should find the website of the British Election Study helpful. The US crime data were obtained from the website provided by the FBI. On behalf of researchers everywhere, I would like to thank these entities for making their data so easy to re-analyze.

GRAHAM J. G. UPTON

CHAPTER 1

INTRODUCTION

This chapter introduces basic statistical ideas and terminology in what the author hopes is a suitably concise fashion. Many readers will be able to turn to Chapter 2 without further ado!

1.1 WHAT ARE CATEGORICAL DATA?

Categorical data are the observed values of variables such as the color of a book, a person's religion, gender, political preference, social class, etc. In short, any variable other than a *continuous variable* (such as length, weight, time, distance, etc.).

If the categories have no obvious order (e.g., Red, Yellow, White, Blue) then the variable is described as a *nominal variable*. If the categories have an obvious order (e.g., Small, Medium, Large) then the variable is described as an *ordinal variable*. In the latter case the categories may relate to an underlying continuous variable where the precise value is unrecorded, or where it simplifies matters to replace the measurement by the relevant category. For example, while an individual's age may be known, it may suffice to record it as belonging to one of the categories "Under 18," "Between 18 and 65," "Over 65."

If a variable has just two categories, then it is a *binary variable* and whether or not the categories are ordered has no effect on the ensuing analysis.

Categorical Data Analysis by Example, First Edition. Graham J. G. Upton.
© 2017 John Wiley & Sons, Inc. Published 2017 by John Wiley & Sons, Inc.

1.2 A TYPICAL DATA SET

The basic data with which we are concerned are *counts*, also called *frequencies*. Such data occur naturally when we summarize the answers to questions in a survey such as that in Table 1.1.

TABLE 1.1 Hypothetical sports preference survey

Sports preference questionnaire
(A) Are you:- Male ☐   Female ☐?
(B) Are you:- Aged 45 or under ☐   Aged over 45 ☐?
(C) Do you:- Prefer golf to tennis ☐   Prefer tennis to golf ☐?

The people answering this (fictitious) survey will be classified by each of the three characteristics: gender, age, and sport preference. Suppose that the 400 replies were as given in Table 1.2 which shows that males prefer golf to tennis (142 out of 194 is 73%) whereas females prefer tennis to golf (161 out of 206 is 78%). However, there is a lot of other information available. For example:

- There are more replies from females than males.
- There are more tennis lovers than golf lovers.
- Amongst males, the proportion preferring golf to tennis is greater amongst those aged over 45 (78/102 is 76%) than those aged 45 or under (64/92 is 70%).

This book is concerned with models that can reveal all of these subtleties simultaneously.

TABLE 1.2 Results of sports preference survey

Category of response	Frequency
Male, aged 45 or under, prefers golf to tennis	64
Male, aged 45 or under, prefers tennis to golf	28
Male, aged over 45, prefers golf to tennis	78
Male, aged over 45, prefers tennis to golf	24
Female, aged 45 or under, prefers golf to tennis	22
Female, aged 45 or under, prefers tennis to golf	86
Female, aged over 45, prefers golf to tennis	23
Female, aged over 45, prefers tennis to golf	75

1.3 VISUALIZATION AND CROSS-TABULATION

While Table 1.2 certainly summarizes the results, it does so in a clumsily long-winded fashion. We need a more succinct alternative, which is provided in Table 1.3.

TABLE 1.3   Presentation of survey results by gender

Male				Female			
Sport	45 and under	Over 45	Total	Sport	45 and under	Over 45	Total
Tennis	28	24	52	Tennis	86	75	161
Golf	64	78	142	Golf	22	23	45
Total	92	102	194	Total	108	98	206

A table of this type is referred to as a *contingency table*—in this case it is (in effect) a three-dimensional contingency table. The locations in the body of the table are referred to as the *cells* of the table. Note that the table can be presented in several different ways. One alternative is Table 1.4.

In this example, the problem is that the page of a book is two-dimensional, whereas, with its three classifying variables, the data set is essentially three-dimensional, as Figure 1.1 indicates. Each face of the diagram contains information about the 2×2 category combinations for two variables for some particular category of the third variable.

With a small table and just three variables, a diagram is feasible, as Figure 1.1 illustrates. In general, however, there will be too many variables and too many categories for this to be a useful approach.

TABLE 1.4   Presentation of survey results by sport preference

Prefers tennis				Prefers golf			
Gender	45 and under	Over 45	Total	Gender	45 and under	Over 45	Total
Female	86	75	161	Female	22	23	45
Male	28	24	52	Male	64	78	142
Total	114	99	213	Total	86	101	187

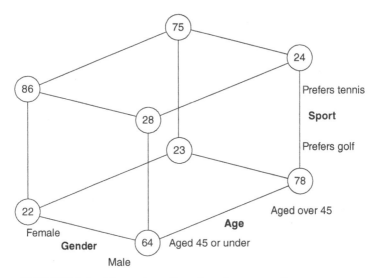

FIGURE 1.1   Illustration of results of sports preference survey.

1.4 SAMPLES, POPULATIONS, AND RANDOM VARIATION

Suppose we repeat the survey of sport preferences, interviewing a second group of 100 individuals and obtaining the results summarized in Table 1.5.

As one would expect, the results are very similar to those from the first survey, but they are not identical. All the principal characteristics (for example, the preference of females for tennis and males for golf) are again present, but there are slight variations because these are the replies from a different set of people. Each person has individual reasons for their reply and we cannot possibly expect to perfectly predict any individual reply since there can be thousands of contributing factors influencing a person's preference. Instead we attribute the differences to *random variation*.

TABLE 1.5   The results of a second survey

	Prefers tennis				Prefers golf		
Gender	45 and under	Over 45	Total	Gender	45 and under	Over 45	Total
Female	81	76	157	Female	16	24	40
Male	26	34	60	Male	62	81	143
Total	107	110	217	Total	78	105	183

Of course, if one survey was of spectators leaving a grand slam tennis tournament, whilst the second survey was of spectators at an open golf tournament, then the results would be very different! These would be *samples* from very different *populations*. Both samples may give entirely fair results for their own specialized populations, with the differences in the sample results reflecting the differences in the populations.

Our purpose in this book is to find succinct models that adequately describe the populations from which samples like these have been drawn. An effective model will use relatively few parameters to describe a much larger group of counts.

1.5 PROPORTION, PROBABILITY, AND CONDITIONAL PROBABILITY

Between them, Tables 1.4 and 1.5 summarized the sporting preferences of 800 individuals. The information was collected one individual at a time, so it would have been possible to keep track of the counts in the eight categories as they accumulated. The results might have been as shown in Table 1.6.

As the sample size increases, so the observed proportions, which are initially very variable, becomes less variable. Each proportion slowly converges on its limiting value, the *population probability*. The difference between columns three and five is that the former is converging on the probability of randomly selecting a particular type of individual from the whole population while the latter is converging on the *conditional probability* of selecting the individual from the relevant subpopulation (males aged over 40).

TABLE 1.6   **The accumulating results from the two surveys**

Sample size	Number of males over 40 who prefer golf	Proportion of sample that are males aged over 40 and prefer golf	Number of males	Proportion of males aged over 40 who prefer golf
10	3	0.300	6	0.500
20	5	0.250	11	0.455
50	8	0.160	25	0.320
100	22	0.220	51	0.431
200	41	0.205	98	0.418
400	78	0.195	194	0.402
800	159	0.199	397	0.401

1.6 PROBABILITY DISTRIBUTIONS

In this section, we very briefly introduce the distributions that are directly relevant to the remainder of the book. A variable is described as being a *discrete variable* if it can only take one of a finite set of values. The probability of any particular value is given by the *probability function*, P.

By contrast, a *continuous variable* can take any value in one or more possible ranges. For a continuous random variable the probability of a value in the interval (a, b) is given by integration of a function f (the so-called *probability density function*) over that interval.

1.6.1 The Binomial Distribution

The binomial distribution is a discrete distribution that is relevant when a variable has just two categories (e.g., Male and Female). If the probability of a randomly chosen individual has probability p of being male, then the probability that a random sample of n individuals contains r males is given by

$$P(r) = \begin{cases} \binom{n}{r} p^r (1-p)^{n-r} & r = 0, 1, \ldots, n, \\ 0 & \text{otherwise,} \end{cases} \tag{1.1}$$

where

$$\binom{n}{r} = \frac{n!}{r!(n-r)!},$$

and

$$r! = r \times (r-1) \times (r-2) \times \cdots \times 2 \times 1.$$

A random variable having such a distribution has *mean* (the average value) np and *variance* (the usual measure of variability) $np(1-p)$. When p is very small and n is large—which is often the case in the context of contingency tables—then the distribution will be closely approximated by a Poisson distribution (Section 1.6.3) with the same mean. When n is large, a normal distribution (Section 1.6.4) also provides a good approximation.

This distribution underlies the logistic regression models discussed in Chapters 7–9.

1.6.2 The Multinomial Distribution

This is the extension of the binomial to the case where there are more than two categories. Suppose, for example, that a mail delivery company classifies packages as being either Small, Medium, and Large, with the proportions falling in these classes being p, q, and $1 - p - q$, respectively. The probability that a random sample of n packages includes r Small packages, s Medium packages, and $(n - r - s)$ Large packages is

$$\frac{n!}{r!s!(n - r - s)!} p^r q^s (1 - p - q)^{n-r-s} \quad \text{where} \quad 0 \le r \le n; \quad 0 \le s \le (n - r).$$

This distribution underlies the models discussed in Chapter 10.

1.6.3 The Poisson Distribution

Suppose that the probability of an individual having a particular characteristic is p, independently, for each of a large number of individuals. In a random sample of n individuals, the probability that exactly r will have the characteristic, is given by Equation (1.1). However, if p (or $1 - p$) is small and n is large, then that binomial probability is well approximated by

$$P(r) = \begin{cases} \dfrac{\mu^r}{r!} e^{-\mu} & r = 0, 1, \dots, \\ 0 & \text{otherwise,} \end{cases} \tag{1.2}$$

where e is the *exponential function* ($=2.71828...$) and $\mu = np$. A random variable with distribution given by Equation (1.2) is said to have a Poisson distribution with *parameter* (a value determining the shape of the distribution) μ. Such a random variable has both mean and variance equal to μ.

This distribution underlies the log-linear models discussed in Chapters 11–16.

1.6.4 The Normal Distribution

The normal distribution (known by engineers as the *Gaussian distribution*) is the most familiar example of a continuous distribution.

If X is a normal random variable with mean μ and variance σ^2, then X has probability density function given by

$$f(x) = \frac{1}{\sigma\sqrt{2\pi}} e^{-(x-\mu)^2/2\sigma^2}. \tag{1.3}$$

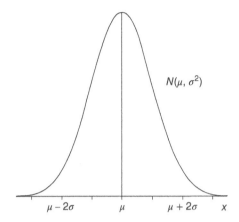

FIGURE 1.2 A normal distribution, with mean μ and variance σ^2.

The density function is illustrated in Figure 1.2. In the case where $\mu = 0$ and $\sigma^2 = 1$, the distribution is referred to as the *standard normal distribution*. Any tables of the normal distribution will be referring to this distribution.

Figure 1.2 shows that most (actually, about 95%) of observations on a random variable lie within about two *standard deviations* (actually 1.96σ) of the mean, with only about three observations in a thousand having values that differ by more than three standard deviations from the mean. The *standard deviation* is the square root of the variance.

1.6.4.1 *The Central Limit Theorem* An informal statement of this theorem is

> A random variable that can be expressed as the sum of a large number of "component" variables which are independent of one another, but all have the same distribution, will have an approximate normal distribution.

The theorem goes a long way to explaining why the normal distribution is so frequently found, and why it can be used as an approximation to other distributions.

1.6.5 The Chi-Squared (χ^2) Distribution

A chi-squared distribution is a continuous distribution with a single parameter known as the *degrees of freedom* (often abbreviated as d.f.). Denoting the value of this parameter by ν, we write that a random variable has a χ^2_ν-distribution. The χ^2 distribution is related to the normal distribution since, if Z has a standard normal distribution, then Z^2 has a χ^2_1-distribution.

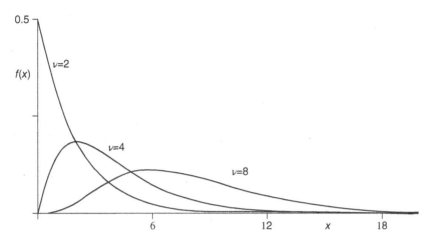

FIGURE 1.3   Chi-squared distributions with 2, 4, and 8 degrees of freedom.

Figure 1.3 gives an idea of what the probability density functions of chi-squared distributions look like. For small values of v the distribution is notably skewed (for $v > 2$, the mode is at $v - 2$). A chi-squared random variable has mean v and variance $2v$.

A very useful property of chi-squared random variables is their additivity: if U and V are independent random variables having, respectively χ^2_u- and χ^2_v- distributions, then their sum, $U + V$, has a χ^2_{u+v} distribution. This is known as the *additive* property of χ^2 distributions.

Perhaps more importantly, if W has a χ^2_w-distribution then it will always be possible to find w independent random variables $(W_1, W_2,..., W_w)$ for which $W = W_1 + W_2 + \cdots + W_w$, with each of $W_1, W_2,..., W_w$ having χ^2_1- distributions. We will make considerable use of this type of result in the analysis of contingency tables.

1.7 *THE LIKELIHOOD

Suppose that n observations, $x_1, x_2,..., x_n$, are taken on the random variable, X. The likelihood, L, is the product of the corresponding probability functions (in the case of a discrete distribution) or probability density functions (in the case of a continuous distribution):

$$L = P(x_1) \times P(x_2) \times \cdots \times P(x_n) \quad \text{or} \quad L = f(x_1) \times f(x_2) \times \cdots \times f(x_n) \quad (1.4)$$

In either case the likelihood is proportional to the probability that a future set of n observations have precisely the values observed in the current set. In most cases, while the form of the distribution of X may be known (e.g., binomial, Poisson), the value of that distribution's parameter (e.g., p, μ) will not be known. Since the observed values have indeed occurred, a logical choice for any unknown parameter would be that value that maximizes the probability of reoccurrence of $x_1, x_2, \ldots, x_n$; this is the principle of *maximum likelihood*.

Suppose that there are r distinct values ($v_1, v_2, \ldots, v_r$) amongst the n observations, with the value v_i occurring on f_i occasions (so $\sum f_i = n$). The distribution that maximizes L is the discrete distribution that precisely describes what has been observed:

$$P(X = x) = \begin{cases} f_i/n & \text{for } x = v_i, \quad i = 1, 2, \ldots, r, \\ 0 & \text{otherwise.} \end{cases} \tag{1.5}$$

CHAPTER 2

ESTIMATION AND INFERENCE FOR CATEGORICAL DATA

This chapter introduces measures of goodness-of-fit and methods for performing tests of hypotheses concerning binomial probabilities. It introduces mid-P and illustrates the difficulties attached to the construction of confidence intervals for binomial probabilities.

2.1 GOODNESS OF FIT

Throughout this book we will be concerned with using models to explain the observed variations in the frequencies of occurrence of various outcomes. In this chapter, we are concerned with the frequencies of occurrence of the categories of a single variable. We begin by introducing two general-purpose goodness-of-fit tests.

2.1.1 Pearson's X^2 Goodness-of-Fit Statistic

Suppose that we have an observed frequency of 20. If the model in question suggests an expected frequency of 20, then we will be delighted (and surprised!). If the expected frequency is 21, then we would not be displeased, but if the expected frequency was 30 then we might be distinctly disappointed. Thus, denoting the observed frequency by f and that expected from the model by e, we observe that the size of $(f - e)$ is relevant.

Categorical Data Analysis by Example, First Edition. Graham J. G. Upton.
© 2017 John Wiley & Sons, Inc. Published 2017 by John Wiley & Sons, Inc.

Now suppose the observed frequency is 220 and the estimate is 230. The value of $(f - e)$ is the same as before, but the difference seems less important because the size of the error is small relative to the size of the variable being measured. This suggests that the proportional error $(f - e)/e$ is also relevant. Both $(f - e)$ and $(f - e)/e$ are embodied in Pearson's X^2 statistic:

$$X^2 = \sum(f - e) \times \frac{(f - e)}{e} = \sum \frac{(f - e)^2}{e}, \qquad (2.1)$$

where the summation is over all categories of interest. The test statistic has an approximate χ^2-distribution and, for this reason, the test is often familiarly referred to as the *chi-squared test*. The test statistic was introduced by the English statistician and biometrician Karl Pearson in 1900.

Example 2.1   Car colors

It is claimed that 25% of cars are red, 30% are white, and the remainder are other colors. A survey of the cars in a randomly chosen car park finds the the results summarized in Table 2.1.

TABLE 2.1   Colors of cars in a car park

Red	White	Others	Total
39	42	119	200

With 200 cars in total, these counts are rather different from those claimed, which would be 50, 60, and 90. To test whether they are significantly different, we calculate

$$X^2 = \frac{(39 - 50)^2}{50} + \frac{(42 - 60)^2}{60} + \frac{(119 - 90)^2}{90} = 2.42 + 5.40 + 9.34 = 17.16.$$

In this case a χ_2^2-distribution is appropriate. The upper 0.1% point of this distribution is 13.82; since 17.16 is much greater than this, we would conclude that the claim concerning the percentages of car colors was incorrect.

2.1.2   *The Link Between X^2 and the Poisson and χ^2-Distributions

Suppose that $y_1, y_2, \ldots, y_n$ are n independent observations from Poisson distributions, with means $\mu_1, \mu_2, \ldots, \mu_n$, respectively. Since a Poisson distribution has its variance equal to its mean,

$$z_i = \frac{y_i - \mu_i}{\sqrt{\mu_i}}$$

will be an observation from a distribution with mean zero and variance one. If μ_i is large, then the normal approximation to a Poisson distribution is relevant and z_i will approximately be an observation from a standard normal distribution. Since the square of a standard normal random variable has a χ_1^2-distribution, that will be the approximate distribution of z_i^2. Since the sum of independent chi-squared random variables is a chi-squared random variable having degrees of freedom equal to the sum of the degrees of freedom of the component variables, we find that

$$\sum_i z_i^2 = \sum \frac{(y_i - \mu_i)^2}{\mu_i}$$

has a chi-squared distribution with n degrees of freedom.

There is just one crucial difference between $\sum z^2$ and X^2: in the former the means are known, whereas for the latter the means are estimated from the data. The estimation process imposes a linear constraint, since the total of the e-values is equal to the total of the f-values. Any linear constraint reduces the number of degrees of freedom by one. In Example 2.1, since there were three categories, there were $(3 - 1) = 2$ degrees of freedom.

2.1.3 The Likelihood-Ratio Goodness-of-Fit Statistic, G^2

Apparently very different to X^2, but actually closely related, is the likelihood-ratio statistic G^2, given by

$$G^2 = 2 \sum f \ln \left(\frac{f}{e} \right), \tag{2.2}$$

where ln is the natural logarithm alternatively denoted as $\log_e$. This statistic compares the maximized likelihood according to the model under test, with the maximum possible likelihood for the given data (Section 1.7).

If the hypothesis under test is correct, then the values of G^2 and X^2 will be very similar. Of the two tests, X^2 is easier to understand, and the individual contributions in the sum provide pointers to the causes of any lack of fit. However, G^2 has the useful property that, when comparing nested models, the more complex model cannot have the larger G^2 value. For this reason the values of both X^2 and G^2 are often reported.

Example 2.1   Car colors (*continued*)

Returning to the data given in Example 2.1, we now calculate

$$2 \left\{ 39 \ln \left(\frac{39}{50} \right) + 42 \ln \left(\frac{42}{60} \right) + 119 \ln \left(\frac{119}{90} \right) \right\} = 2 \times (-9.69 - 14.98 + 33.24)$$

$$= 17.13.$$

As foreshadowed, the value of G^2 is indeed very similar to that of $X^2 (= 17.16)$ and the conclusion of the test is the same.

2.1.4 *Why the G^2 and X^2 Statistics Usually Have Similar Values

This section is included to satisfy those with an inquiring mind! For any model that provides a tolerable fit to the data, the values of the observed and expected frequencies will be similar, so that f/e will be reasonably close to 1. We can employ a standard mathematical "trick" and write $f = e + (f - e)$, so that

$$\frac{f}{e} = 1 + \frac{f-e}{e},$$

and

$$G^2 = 2\sum f \ln\left(\frac{f}{e}\right) = 2\sum\{e + (f - e)\}\ln\left(1 + \frac{f-e}{e}\right).$$

For small δ, the Maclaurin series expansion of $\ln(1 + \delta)$ is

$$\delta - \frac{\delta^2}{2} + \frac{\delta^3}{3} - \cdots.$$

If the model is a reasonable fit, then $(f - e)/e$ will be small and so

$$G^2 = 2\sum\{e + (f - e)\}\left\{\frac{f-e}{e} - \frac{(f-e)^2}{2e^2} + \cdots\right\}$$

$$= 2\sum\left\{\left((f - e) - \frac{(f-e)^2}{2e} + \cdots\right) + \left(\frac{(f-e)^2}{e} - \cdots\right)\right\}.$$

But $\sum(f - e) = 0$ since $\sum f = \sum e$, and so

$$G^2 \approx 0 - \sum\frac{(f-e)^2}{e} + 2\sum\frac{(f-e)^2}{e} = \sum\frac{(f-e)^2}{e} = X^2.$$

2.2 HYPOTHESIS TESTS FOR A BINOMIAL PROPORTION (LARGE SAMPLE)

If a categorical variable has several categories then it is natural to inquire about the probability of occurrence for each category. If a particular category occurs on r occasions in n trials, then the unbiased estimate of p, the probability of occurrence of that category, is given by $\hat{p} = r/n$.

We are interested in testing the hypothesis H_0, that the population probability is p_0 against the alternative H_1, that H_0 is false. With n observations, under the null hypothesis, the expected number is np_0 and the variance is $np_0(1 - p_0)$.

2.2.1 The Normal Score Test

If n is large, then a normal approximation should be reasonable so that we can treat r as an observation from a normal distribution with mean np_0 and variance $np_0(1 - p_0)$. The natural test is therefore based on the value of z given by

$$z = \frac{r - np_0}{\sqrt{np_0(1 - p_0)}}. \tag{2.3}$$

The value of z is compared with the distribution function of a standard normal distribution to determine the test outcome which depends on the alternative hypothesis (one-sided or two-sided) and the chosen significance level.

2.2.2 *Link to Pearson's X^2 Goodness-of-Fit Test

Goodness-of-fit tests compare observed frequencies with those expected according to the model under test. In the binomial context, there are two categories with observed frequencies r and $n - r$ and expected frequencies np_0 and $n(1 - p_0)$. The X^2 goodness-of-fit statistic is therefore given by

$$
\begin{aligned}
X^2 &= \frac{(r - np_0)^2}{np_0} + \frac{\{(n - r) - n(1 - p_0)\}^2}{n(1 - p_0)} = \frac{(r - np_0)^2}{np_0} + \frac{(r - np_0)^2}{n(1 - p_0)} \\
&= \frac{(r - np_0)^2}{n} \left\{ \frac{1}{p_0} + \frac{1}{1 - p_0} \right\} \\
&= \frac{(r - np_0)^2}{np_0(1 - p_0)} \\
&= z^2.
\end{aligned}
$$

The normal score test and the X^2-test are therefore equivalent and thus X^2 has an approximate χ_1^2-distribution.

2.2.3 G^2 for a Binomial Proportion

In this application of the likelihood-ratio test, G^2 is given by:

$$G^2 = 2r \ln \left(\frac{r}{np_0} \right) + 2(n - r) \ln \left(\frac{n - r}{n(1 - p_0)} \right). \tag{2.4}$$

It has an approximate χ_1^2-distribution.

Example 2.2   Colors of sweet peas

A theory suggests that the probability of a sweet pea having red flowers is 0.25. In a random sample of 60 sweet peas, 12 have red flowers. Does this result provides significant evidence that the theory is incorrect?

The theoretical proportion is $p = 0.25$. Hence $n = 60$, $r = 12$, $n - r = 48$, $np = 15$, and $n(1 - p) = 45$. Thus

$$z = \frac{15 - 12}{\sqrt{60 \times 0.25 \times 0.75}} = \frac{3}{\sqrt{11.25}} = 0.8944,$$

$$X^2 = 0.8,$$

$$G^2 = 24 \ln(12/15) + 96 \ln(48/45) = 0.840.$$

The tail probability for z (0.186) is half that for X^2 (0.371) because the latter refers to two tails. The values of X^2 and G^2 are, as expected, very similar. All the tests find the observed outcome to be consistent with theory.

2.3 HYPOTHESIS TESTS FOR A BINOMIAL PROPORTION (SMALL SAMPLE)

The problem with any discrete random variable is that probability occurs in chunks! In place of a smoothly increasing distribution function, we have a step function, so that only rarely will there be a value of x for which $P(X \geq x)$ is exactly some pre-specified value of α. This contrasts with the case of a continuous variable, where it is almost always possible to find a precise value of x that satisfies $P(X > x) = \alpha$ for any specified α.

Another difficulty with any discrete variable is that, if $P(X = x) > 0$, then

$$P(X \geq x) + P(X \leq x) = 1 + P(X = x),$$

and the sum is therefore greater than 1. For this reason, Lancaster (1949) suggested that, for a discrete variable, rather than using $P(X \geq x)$, one should use

$$P_{Mid}(x) = \frac{1}{2} P(X = x) + P(X > x). \tag{2.5}$$

This is called the *mid-P value*; there is a corresponding definition for the opposite tail, so that the two mid-P values do sum to 1.

2.3.1 One-Tailed Hypothesis Test

Suppose that we are concerned with the unknown value of p, the probability that an outcome is a "success." We wish to compare the null hypothesis

$H_0 : p = \frac{1}{4}$ with the alternative hypothesis $H_1 : p > \frac{1}{4}$, using a sample of size 7 and a 5% tail probability.

Before we carry out the seven experiments, we need to establish our test procedure. There are two reasons for this:

1. We need to be sure that a sample of this size can provide useful information concerning the two hypotheses. If it cannot, then we need to use a larger sample.
2. There are eight possible outcomes (from 0 successes to 7 successes). By setting out our procedure *before* studying the actual results of the experiments, we are guarding against biasing the conclusions to make the outcome fit our preconceptions.

The binomial distribution with $n = 7$ and $p = 0.25$ is as follows:

x	0	1	2	3	4	5	6	7
$P(X = x)$	0.1335	0.3115	0.3115	0.1730	0.0577	0.0115	0.0013	0.0001
$P(X \geq x)$	1.0000	0.8665	0.5551	0.2436	0.0706	0.0129	0.0013	0.0001
$P_{Mid}(x)$		0.9333	0.7108	0.3993	0.1571	0.0417	0.0071	0.0007	0.0000

We intended to perform a significance test at the 5% level, but this is impossible! We can test at the 1.3% level (by rejecting if $X \geq 5$) or at the 7.1% level, (by rejecting if $X \geq 5$), but not at exactly 5%. We need a rule to decide which to use.

Here are two possible rules (for an upper-tail test):

1. Choose the smallest value of X for which the significance level does not exceed the target level.
2. Choose the smallest value of X for which the mid-P value does not exceed the target significance level.

Because the first rule guarantees that the significance level is never greater than that required, on average it will be less. The second rule uses mid-P defined by Equation (2.5). Because of that definition, whilst the significance level used under this rule will sometimes be greater than the target level, on average it will equal the target.

For the case tabulated, with the target of 5%, the conservative rule would lead to rejection only if $X \geq 5$ (significance level 1.3%), whereas, with the mid-P rule, rejection would also occur if $X = 4$ (significance level 7.1%).

2.3.2 Two-Tailed Hypothesis Tests

For discrete random variables, we regard a two-tailed significance test at the $\alpha\%$ level as the union of two one-tailed significance tests, each at the $\frac{1}{2}\alpha\%$ level.

Example 2.3   Two-tailed example

Suppose we have a sample of just six observations, with the hypotheses of interest being $H_0 : p = 0.4$ and $H_1 : p \neq 0.4$. We wish to perform a significance test at a level close to 5%. According to the null hypothesis the complete distribution is as follows:

x	0	1	2	3	4	5	6
$P(X = x)$	0.0467	0.1866	0.3110	0.2765	0.1382	0.0369	0.0041
$P(X \geq x)$	1.0000	0.9533	0.7667	0.4557	0.1792	0.0410	0.0041
$P_{Mid}(x)$ (upper tail)	0.9767	0.8600	0.6112	0.3174	0.1101	0.0225	0.0020
$P(X \leq x)$	0.0467	0.2333	0.5443	0.8208	0.9590	0.9959	1.0000
$P_{Mid}(x)$ (lower tail)	0.0233	0.1400	0.3888	0.6826	0.8899	0.9775	0.9980

Notice that the two mid-P values sum to 1 (as they must). We aim for 2.5% in each tail. For the upper tail the appropriate value is $x = 5$ (since 0.0225 is just less than 0.025). In the lower tail the appropriate x-value is 0 (since 0.0233 is also just less than 0.25). The test procedure is therefore to accept the null hypothesis unless there are 0, 5, or 6 successes. The associated significance level is $0.0467 + 0.0369 + 0.0041 = 8.77\%$. This is much greater than the intended 5%, but, in other cases, the significance level achieved will be less than 5%. On average it will balance out.

2.4 INTERVAL ESTIMATES FOR A BINOMIAL PROPORTION

If the outcome of a two-sided hypothesis test is that the sample proportion is found to be consistent with the the population proportion being equal to p_0, then it follows that p_0 is consistent with the sample proportion. Thus, a confidence interval for a binomial proportion is provided by the range of values of p_0 for which this is the case. In a sense, hypothesis tests and confidence intervals are therefore two sides of the same coin.

Since a confidence interval provides the results for an infinite number of hypothesis tests, it is more informative. However, as will be seen, the determination of a confidence interval is not as straightforward as determining the outcome of a hypothesis test. The methods listed below are either those most often used or those that appear (to the author at the time of writing) to be the most accurate. There have been several other variants put forward in the past 20 years, as a result of the ability of modern computers to undertake extensive calculations.

2.4.1 Laplace's Method

A binomial distribution with parameters n and p has mean np and variance $np(1 - p)$. When p is unknown it is estimated by $\hat{p} = r/n$, where r is the number of successes in the n trials. If p is not near 0 or 1, one might anticipate that $p(1 - p)$ would be closely approximated by $\hat{p}(1 - \hat{p})$. This reasoning led the French mathematician Laplace to suggest the interval:

$$\frac{r}{n} \pm z_0 \sqrt{\frac{1}{n}\frac{r}{n}\left(1 - \frac{r}{n}\right)}, \tag{2.6}$$

where z_0 is the appropriate critical value from a standard normal distribution.

Unfortunately, using Equation (2.6), the actual size of the confidence interval can be much smaller than its intended value. For the case $n = 12$, with the true value of p being 0.5, Brown, Cai, and DasGupta (2001) report that the average size of the "95%" interval is little bigger than 85%. The procedure gets worse as the true value of p diverges from 0.5 (since the chance of $r = 0$ or $r = n$ increases, and those values would lead to an interval of zero width).

A surprising feature (that results from the discreteness of the binomial distribution) is that an increase in sample size need not result in an improvement in accuracy (see Brown et al. (2001) for details). Although commonly cited in introductory texts, the method cannot be recommended.

2.4.2 Wilson's Method

Suppose that z_c is a critical value of the normal score test (Equation 2.3), in the sense that any absolute values greater than z_c would lead to rejection of the null hypothesis. For example, for a two-sided 5% test, $z_c = 1.96$. We are interested in finding the values of p_0 that lead to this value. This requires the solution of the quadratic equation

$$z_0^2 \times np_0(1 - p_0) = (r - np_0)^2,$$

which has solutions

$$\left\{ \left(2r + z_0^2\right) \pm z_0 \sqrt{z_0^2 + 4r - 4r^2/n} \right\} / 2\left(n + z_0^2\right). \qquad (2.7)$$

This interval was first discussed by Wilson (1927).

2.4.3 The Agresti–Coull Method

A closely related but simpler alternative suggested by Agresti and Coull (1998) is

$$\tilde{p} \pm z_o \sqrt{\frac{1}{n}\tilde{p}(1 - \tilde{p})}, \quad \text{where} \quad \tilde{p} = \left(2r + z_0^2\right)/2\left(n + z_0^2\right). \qquad (2.8)$$

Since, at the 95% level, $z_0^2 = 3.84 \approx 4$, this 95% confidence interval effectively works with a revised estimate of the population proportion that adds two successes and two failures to those observed.

Example 2.4   Proportion smoking

In a random sample of 250 adults, 50 claim to have never smoked. The estimate of the population proportion is therefore 0.2. We now determine 95% confidence intervals for this proportion using the methods of the last sections.

The two-sided 95% critical value from a normal distribution is 1.96, so Laplace's method gives the interval $0.2 \pm 1.96\sqrt{0.2 \times 0.8/250} =$ $(0.150, 0.250)$. For the Agresti–Coull method $\tilde{p} = 0.205$ and the resulting interval is $(0.155, 0.255)$. Wilson's method also focuses on $\tilde{p}$, giving $(0.155, 0.254)$. All three estimates are reassuringly similar.

Suppose that in the same sample just five claimed to have smoked a cigar. This time, to emphasize the problems and differences that can exist, we calculate 99% confidence intervals. The results are as follows: Laplace $(-0.003, 0.043)$, Agresti–Coull $(-0.004, 0.061)$, Wilson $(-0.007, 0.058)$. The differences in the upper bounds are quite marked and all three give impossibly negative lower bounds.

2.4.4 Small Samples and Exact Calculations

**2.4.4.1 *Clopper–Pearson Method*   For small samples, Clopper and Pearson (1934) suggested treating the two tails separately. Denoting the lower and upper bounds by p_L and p_U, for a $(100 - \alpha)\%$ confidence interval, these would be the values satisfying:

$$\sum_{k=r}^{n} \binom{n}{k} p_L^k (1 - p_L)^{n-k} = \frac{1}{2}\alpha \quad \text{and} \quad \sum_{k=0}^{r} \binom{n}{k} p_U^k (1 - p_U)^{n-k} = \frac{1}{2}\alpha.$$

The method is referred to as "exact" because it uses the binomial distribution itself, rather than an approximation. As such it cannot lead to bounds that lie outside the feasible region (0, 1).

However, just as Laplace's method leads to overly narrow confidence intervals, so the Clopper–Pearson approach leads to overly wide confidence intervals, with the true width (the *cover*) of a Clopper–Pearson interval being greater than its nominal value.

2.4.4.2 *Agresti–Gottard Method*   Agresti and Gottard (2007) suggested a variant of the Clopper–Pearson approach that gives intervals that on average are superior in the sense that their average cover is closer to the nominal $100(1 - \alpha)\%$ value. The variant makes use of mid-P (Equation 2.5):

$$\sum_{k=(r+1)}^{n} \binom{n}{k} p_L^k (1 - p_L)^{n-k} + \frac{1}{2} \binom{n}{r} p_L^r (1 - p_L)^{n-r} = \frac{1}{2}\alpha, \qquad (2.9)$$

$$\sum_{k=0}^{r-1} \binom{n}{k} p_U^k (1 - p_U)^{n-k} + \frac{1}{2} \binom{n}{r} p_U^r (1 - p_U)^{n-r} = \frac{1}{2}\alpha, \qquad (2.10)$$

with obvious adjustments if $r = 0$ or $r = n$. A review of the use of mid-P with confidence intervals is provided by Berry and Armitage (1995).

Figure 2.1 shows the surprising effect of discreteness on the actual average size of the "95%" mid-P confidence intervals constructed for the case $n = 50$. On the x-axis is the true value of the population parameter, evaluated between 0.01 and 0.99, in steps of size 0.01. The values vary between 0.926 and 0.986 with mean 0.954. The corresponding plot for the Clopper–Pearson procedure and, indeed, for any other alternative procedure will be similar. Similar results hold for the large-sample methods: the best advice would be to treat any intervals as providing little more than an indication of the precision of an estimate.

Example 2.4   Proportion smoking (*continued*)

Returning to the smoking example we can use the `binom.midp` function which is part of the `binomSamSize` library in R to calculate confidence intervals that use Equations (2.9) and (2.10). The results are (0.154, 0.253) as the

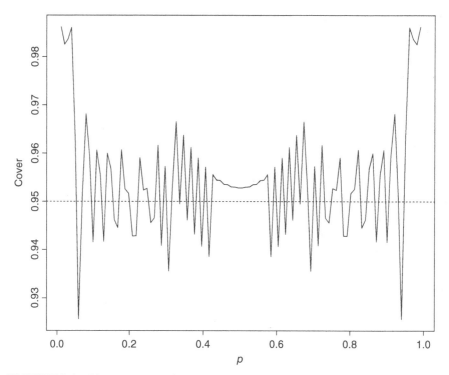

FIGURE 2.1 The average width of a "95%" confidence interval for p varies with the actual p-value. The results shown are using the recommended mid-P interval for the case $n = 50$.

95% interval for the proportion who had never smoked and (0.005, 0.053) as the 99% interval for those who had smoked a cigar. For the former, the results are in good agreement with the large sample approximate methods. For the cigar smokers, the method provides a reassuringly positive lower bound.

R code

```
library(binomSamSize);
binom.midp(50,250,0.95);
binom.midp(5,250,0.99)
```

REFERENCES

Agresti, A., and Coull, B. A. (1998) Approximate is better than exact for interval estimation of binomial parameters. *Am. Stat.*, **52**, 119–126.

Agresti, A., and Gottard, A. (2007) Nonconservative exact small-sample inference for discrete data. *Comput. Stat. Data. An.*, **51**, 6447–6458.

Berry, G., and Armitage, P. (1995) Mid-P confidence intervals: a brief review. *J. R. Stat. Soc. D*, **44**, 417–423.

Brown, L. D., Cai, T. T., and DasGupta, A. (2001) Interval estimation for a binomial proportion. *Stat. Sci.*, **16**, 101–133.

Clopper, C., and Pearson, E. S. (1934) The use of confidence or fiducial limits illustrated in the case of the binomial. *Biometrika*, **26**, 404–413.

Lancaster, H. O. (1949) The combination of probabilities arising from data in discrete distributions. *Biometrika*, **36**, 370–382.

Pearson, K. (1900) On the criterion that a given system of deviations from the probable in the case of a correlated system of variables is such that it can be reasonably supposed to have arisen from random sampling. *Philos. Mag. Ser. 5*, **50**(302), 157–175.

Wilson, E. B. (1927) Probable inference, the law of succession, and statistical inference. *J. Am. Stat. Assoc.*, **22**, 209–212.

CHAPTER 3

THE 2 × 2 CONTINGENCY TABLE

The 2×2 contingency table consists of just four numbers arranged in two rows with two columns to each row; a very simple arrangement. In part, it is this very simplicity that has been the cause of so much discussion concerning the correct method of analysis. Indeed, there may well be more papers written concerning the analysis of two-by-two contingency tables than there are on any other statistical topic!

3.1 INTRODUCTION

The data consist of "experimental units," classified by the categories to which they belong, for each of two *dichotomous variables* (variables having just two categories). The experimental units may be tangible (e.g., people, universities, countries, buildings) or intangible (e.g., ideas, opinions, outcomes). As an example, suppose that people are cross-classified by gender (male or female) and eye color (brown, not brown). A question of possible interest is whether gender and eye color are independent of one another, or whether they are in

Categorical Data Analysis by Example, First Edition. Graham J. G. Upton.
© 2017 John Wiley & Sons, Inc. Published 2017 by John Wiley & Sons, Inc.

some way associated. For example, is the proportion of males having brown eyes different to the proportion of females having brown eyes?

Here are some possible sampling schemes:

1. Take a random sample of individuals. In the time available, we collect information on N individuals. Of these m happen to be male, $n = (N - m)$ happen to be female, r happen to have brown eyes, and $s = (N - r)$ do not have brown eyes. The full results are summarized in the table:

	Brown eyes	Not brown eyes	Total
Male	a	b	m
Female	c	d	n
	r	s	N

2. Take a random sample of N individuals. It is found that m are male and r have brown eyes. The full results are as previously.

3. Take a random sample of N individuals. It is found that m are male and it is known that r will have a particular attribute (see Example 3.2 for such a situation). The full results are as previously.

4. Take independent random samples of m males and n females. In all it is found that r have brown eyes and $s = (N - r)$ do not have brown eyes. The full results are as previously.

5. Take independent random samples of r individuals who have brown eyes and s individuals who do not have brown eyes. In all it is found that m are male and n are female. The full results are as previously.

6. Continue taking observations until a males with brown eyes have been sampled. The full results are then found to be as previously.

In each case the question of interest is the same and the same results have been found. The only differences relate to which (if any) of the *marginal totals* (m, n, r, s, N) have been fixed by the sampling scheme. Since the question of interest is the same and the same results have been obtained, it might seem perverse to consider applying different tests in different situations. Nevertheless, there has been much debate concerning the extent to which the analysis should be affected by the method by which the data were obtained. This has continued ever since the initial proposal of Fisher (1935), which we now introduce.

3.2 FISHER'S EXACT TEST (FOR INDEPENDENCE)

It is convenient to rewrite the table as follows

	Brown eyes	Not brown eyes	Total
Male	a	$m - a$	m
Female	$r - a$	$n - r + a$	n
	r	s	N

Written in this way it is apparent that this is effectively a single variable situation. As we change the value of a (we could as easily have focused on b, c, or d) so the values of the remaining three cells in the body of the table change accordingly. The range of possible values of a is restricted by the need for each of a, $(m - a)$, $(r - a)$, and $(n - r + a)$ to remain non-negative.

Writing A as the random variable corresponding to the observed value of a, Fisher (1934) showed that

$$P(A = a) = \frac{m!n!r!s!}{N!} \times \frac{1}{a!b!c!d!}. \tag{3.1}$$

The test is often referred to simply as *the exact test*—although, of course, it is not the only test that is exact! For arguments in favor of its general application, see Yates (1984), Cormack and Mantel (1991), and Upton (1992). For an alternative view, see Rice (1988).

Despite its apparent simplicity, precisely how one uses Equation (3.1) is not as straightforward as one might expect, though there is no doubt about the values of the single-tail probabilities $P(A \geq a)$ and $P(A \leq a)$. The difficulty arises with the two-tailed test of the null hypothesis of independence (the most common situation), since it is not obvious how to incorporate probabilities from the tail opposite to that which occurred. Here are some feasible alternatives for the case where the observed value, a, lies in the upper tail of the distribution of A:

1. Determine the sum of all the individual outcome probabilities that are no larger than $P(A = a)$. This approach, suggested by Irwin (1935), is the usual method.
2. Use the smaller of $2P(A \geq a)$ and 1.
3. Determine a^*, the largest value of A for which $P(A \leq a^*) \leq P(A \geq a)$. Use the sum $\{P(A \leq a^*) + P(A \geq a)\}$. Depending on the shape of the

distribution of A, this suggestion by Blaker (2000, 2001) may give a different result to that of Irwin.

4. Use mid-P (Equation 2.5). This implies reducing the probability (calculated using either Irwin's method or Blaker's method) by $\frac{1}{2}P(A = a)$.

Example 3.1   Tail probabilities for a 2 × 2 table

Suppose $m = 4, n = 14, r = 11$, and $s = 7$, with $N = 18$. One of the following tables must occur:

0	4		1	3		2	2		3	1		4	0
11	3		10	4		9	5		8	6		7	7

The corresponding probabilities are

$$\frac{49}{4284} \qquad \frac{539}{4284} \qquad \frac{1617}{4284} \qquad \frac{1617}{4284} \qquad \frac{462}{4284}.$$

The corresponding two-tail probabilities calculated by the various methods are summarized in Table 3.1 for each of the possible procedures.

TABLE 3.1   Two-tail probabilities for a 2 × 2 table, calculated using a variety of rules

Value of a	0	1	2	3	4
Irwin: sum of small probabilities	0.0114	0.2451	1.0000	1.0000	0.1193
Capped double tail	0.0229	0.2745	1.0000	0.9767	0.2157
Blaker: sum of tails	0.0114	0.2451	1.0000	0.6225	0.1193
Mid-P with Irwin	0.0057	0.1822	0.8113	0.8113	0.0654
Mid-P with Blaker	0.0057	0.1822	0.8113	0.4338	0.0654

Although there is a striking difference between the Irwin and Blaker tail probabilities for one outcome here, the reader can be comforted that this is a reasonably uncommon occurrence and any difference between them is unlikely to be as large as in the current case. The use of the doubled single tail does not appear appropriate. The optimal combination may be provided by using Blaker's suggestion together with mid-P.

3.2.1 *Derivation of the Exact Test Formula

Denote the probability of a male having brown eyes by p. According to the null hypothesis (independence of gender and eye color) the probability of a female having brown eyes is also p.

The probability that a of m males have brown eyes is

$$\binom{m}{a} p^a (1-p)^{m-a}.$$

The corresponding probability for the females is

$$\binom{n}{r-a} p^{r-a}(1-p)^{(n-r+a)}.$$

So the (unconditional) probability of the observed outcome is

$$\binom{m}{a} p^a (1-p)^{m-a} \times \binom{n}{r-a} p^{r-a}(1-p)^{(n-r+a)} = \binom{m}{a} \binom{n}{r-a} p^r (1-p)^s.$$

However, we know that there is a total of r brown-eyed people, for which the probability is

$$\sum_a \binom{m}{a} \binom{n}{r-a} p^r (1-p)^s = p^r (1-p)^s \sum_a \binom{m}{a} \binom{n}{r-a}.$$

Comparing the coefficients of x^r in $(1+x)^m (1+x)^n$ and in $(1+x)^{m+n}$ we find that

$$\sum_a \binom{m}{a} \binom{n}{r-a} = \binom{m+n}{r}.$$

Hence, the probability of the observed outcome *conditional* on the marginal totals is

$$\binom{m}{a} \binom{n}{r-a} p^r (1-p)^s \bigg/ p^r (1-p)^s \binom{m+n}{r} = \frac{m!n!r!s!}{a!b!c!d!N!}.$$

3.3 TESTING INDEPENDENCE WITH LARGE CELL FREQUENCIES

When it was introduced (and for decades afterward) use of the exact test was confined to cases where the the cell frequencies were small. Since the validity of the test is unaffected by sample size, and since modern computer programs can handle cell frequencies of any size (either by direct evaluation or by simulation), the exact test is the one to use (probably using Blaker's variant and mid-P) if it is available.

However, for many years the tests in the next subsections were the only ones feasible for large samples. These therefore appear prominently in older texts. We consider them here because it is always convenient to have available a test that can be calculated on the back of an envelope!

3.3.1 Using Pearson's Goodness-of-Fit Test

Suppose that eye color and gender are independent. In our example, since the observed proportion having brown eyes is r/N, the expected number of brown-eyed males is mr/N, and the entire table of expected frequencies is as follows:

	Brown eyes	Not brown eyes	Total
Male	mr/N	ms/N	m
Female	nr/N	ns/N	n
Total	r	s	N

Using Equation (2.1) we have

$$X^2 = \frac{(a - mr/N)^2}{mr/N} + \cdots + \frac{(d - ns/N)^2}{ns/N}.$$

After a bit of algebra, this distinctly cumbersome expression reduces to something simple, namely:

$$X^2 = \frac{N(ad - bc)^2}{mnrs}. \tag{3.2}$$

For the 2×2 table the approximating reference distribution is a χ_1^2-distribution. If the observed value of X^2 exceeds some pre-determined significance point of this distribution then the null hypothesis of independence would be rejected.

3.3.2 The Yates Correction

An equivalent test would compare

$$X = \sqrt{\frac{N}{mnrs}}(ad - bc)$$

with a standard normal distribution. For any given values of m, n, r, and s, there are only a limited number of possible values for $(ad - bc)$, since the values of a, b, c, and d are integers. The true distribution of X is therefore a discrete distribution for which the standard normal is not more than an approximation. Discreteness was, of course, equally true for X^2, but it is easier to see how to improve the continuous approximation when working with X.

Suppose that a is greater than the expected value mr/N. Then $P(A \geq a)$ would be estimated by using $a - \frac{1}{2}$ as the critical value of the approximating continuous variable. The $\frac{1}{2}$ is the standard continuity correction. To preserve the marginal totals, each of the observed counts in the table needs to be adjusted by $\frac{1}{2}$ (two counts upward, and two counts downward). Using the revised values, algebra again rescues an unpromising situation and gives as the continuity-corrected X^2 value:

$$X_c^2 = \frac{N(|ad - bc| - \frac{1}{2}N)^2}{mnrs}. \tag{3.3}$$

This version of the Pearson test was introduced in Yates (1934) and is known as the Yates-corrected chi-squared test. Although the Yates correction appears in most introductory textbooks, the following argument suggests that it should not be used:

- The continuity correction gives tail probabilities using X_c^2 that are very close to those obtained using the exact test (Upton, 1982).
- However, hypothesis tests for discrete distributions should preferably use mid-P.
- The mid-P adjustment involves half the probability of the observed value. This adjustment works in the opposite direction to the continuity correction and effectively cancels it.

The implication is that, if using an approximate test, it is the uncorrected X^2 given by Equation (3.2) that is appropriate and not the corrected test.

Example 3.2 Possible gender bias in the 2005 UK General Election

In the 2005 UK General Election, there were 646 seats contested by 3552 candidates of whom 737 were female. There were 127 females elected. Is there evidence of a candidate's gender affecting that candidate's chance of success?

Arranged as a table, the results are:

	Elected	Not elected	Total
Male	519	2298	2817
Female	127	608	735
Total	646	2906	3552

Note that, since the numbers of male and female candidates and the number that can be successful as a result of the voting are known in advance of election day, this is an example in which *both* sets of marginal totals are fixed. The issue is whether there is any link between gender and success, or whether the probability of success is unaffected by gender (which is the null hypothesis).

The value of X^2 is 0.51, which is certainly not significantly large. We can accept the hypothesis that there is no difference in the success rates of the male and female candidates. Of course, there *is* a distinct bias visible in the table since there are roughly four male candidates for each female candidate, but we would need other data to decide whether that was a consequence of discrimination.

3.4 THE 2 × 2 TABLE IN A MEDICAL CONTEXT

Consider the following 2 × 2 table:

Test	Patient		Total
	has disease	does not have disease	
suggests patient has disease	*a* **True positives** **(TP)**	*b* **False positives** **(FP)**	*m*
does not suggest patient has disease	*c* **False negatives** **(FN)**	*d* **True negatives** **(TN)**	*n*
Total	*r*	*s*	*N*

Here we do not expect independence—quite the reverse! Ideally *b* and *c* would both be zero and the test would be infallible. Sadly that is rarely true.

What will be true (for any respectable test) is that, a/m, the *conditional probability* of a patient having the disease given that that is what the test suggests, will be greater than c/n, the conditional probability of a patient having the disease when the test says that that is not the case.

There are several conditional probabilities that will be of interest to the clinician. Yerushalmy (1947) suggested concentrating attention on the following:

Sensitivity:	The probability that a patient with the disease is correctly diagnosed	$= a/r$.
Specificity:	The probability that a patient without the disease is correctly diagnosed	$= d/s$.

A test that suggested that every patient had the disease would have sensitivity $= 1$, but specificity $= 0$. This led Youden (1950) to suggest combining the two measures to give a single index of the test's performance:

$$J = \text{Sensitivity} + \text{Specificity} - 1.$$

This is now referred to as *Youden's index*. The index is a good measure of the usefulness of a test, but it gives little information concerning the correctness of the diagnosis. After the diagnosis, it is a second pair of conditional probabilities that will be of interest to the patient. These are:

Positive predictive value (PPV):	The probability that a patient diagnosed as having the disease is correctly diagnosed	$= a/m$.
Negative predictive value (NPV):	The probability that a patient diagnosed as not having the disease is correctly diagnosed	$= d/n$.

However, the values of PPV and NPV will be affected by the overall *prevalence* (r/N) of the disease. If every patient had the disease then PPV $= 1$ and NPV $= 0$.

In the very different context of multiple hypothesis tests, Benjamini and Hochberg (1995) described the expected number of null hypotheses that were incorrectly rejected as the *false discovery rate*. In the current notation, with the patient having the disease being equated to the null hypothesis being true, the false discovery rate is the expected value of c/n.

Two further quantities of interest are:

Positive likelihood ratio: Sensitivity/(1−Specificity) $= \dfrac{a/r}{b/s} = as/br.$

Negative likelihood ratio: (1−Sensitivity)/Specificity $= \dfrac{c/r}{d/s} = cs/dr.$

Thus the positive likelihood ratio is the ratio between the probability of a positive test result given the presence of the disease and the probability of a positive test result given the absence of the disease; this should be very large. The negative likelihood ratio is the ratio between the probability of a negative test result given the presence of the disease and the probability of a negative test result given the absence of the disease; this should be less than 1 and close to 0.

Example 3.3 Radiological diagnoses of cancer in Haarlem

Table 3.2 refers to patients screened for breast cancer in a Haarlem hospital during the calendar years 1992 and 1993.

TABLE 3.2 The success of radiological diagnoses of cancer in Haarlem

	Biopsy or follow-up result		
Radiological diagnosis	Carcinoma	No carcinoma	Total
Suspicious or malignant	138	65	203
Normal, benign, or probably benign	12	2799	2811
Total	150	2864	3014

Source: Duijm et al., 1997. Reproduced with permission of Elsevier.

The radiological diagnosis evidently works well giving sensitivity = 138/150 (92.0%), specificity = 2799/2864 (97.7%), positive predictive value = 138/203 (68.0%), negative predictive value = 2799/2811 (99.6%), positive likelihood ratio = 40.5, and negative likelihood ratio = 0.08.

3.5 MEASURING LACK OF INDEPENDENCE (COMPARING PROPORTIONS)

The terms used in medical contexts provided specific information about the efficiency (or otherwise) of a diagnostic procedure for a single group of N individuals. We now consider the case of two distinct groups (of sizes m and n), with the probabilities of obtaining a "success" being p_1 and p_2, respectively. We are concerned with the relative sizes of these probabilities.

3.5.1 Difference of Proportions

With a successes in m trials, the estimated probability of a success is $\hat{p}_1 = a/m$, with b/n being the corresponding estimate for the second sample. The difference between proportions is estimated by

$$\hat{p}_1 - \hat{p}_2 = \frac{a}{m} - \frac{b}{n} = \frac{an - bm}{mn} = \frac{ad - bc}{mn}.$$

The value a is an observation from a binomial distribution with parameters m and p_1. The distribution has mean mp_1 and variance $mp_1(1 - p_1)$. Thus $\hat{p}_1$ is an observation from a distribution with mean p_1 and variance $p_1(1 - p_1)/m$. It follows that $\hat{p}_1 - \hat{p}_2$ is an observation from a distribution with mean $p_1 - p_2$ and variance $p_1(1 - p_1)/m + p_2(1 - p_2)/n$. If m and n are not small, and neither p_1 nor p_2 is close to 0 or 1, then the normal approximation to the binomial distribution can be used. With z_0 denoting the appropriate critical value from a standard normal distribution, an approximate confidence interval is given by

$$(\hat{p}_1 - \hat{p}_2) \pm z_0 \sqrt{\frac{1}{m}\hat{p}_1 (1 - \hat{p}_1) + \frac{1}{n}\hat{p}_2 (1 - \hat{p}_2)}. \qquad (3.4)$$

Example 3.4   Pot decoration in the Admiralty Islands

Table 3.3 presents information on the coloration of pots made by two communities in the Admiralty Islands. One feature was the lips of the pots. Decorated lips were found in 86 of 437 pots produced by the Hus community, and in 169 of 439 pots produced by the Mbuke community. A question of interest is whether there is a significant difference between the two communities concerning the proportions of pots having decorated lips.

TABLE 3.3   Pot decoration in the Admiralty Islands

	Community	
	Hus	Mbuke
Decorated lip	86	169
Undecorated lip	351	270
Total	437	439

Source: Kennedy, 1981. Reproduced with permission of Taylor & Francis Ltd (www.tandfonline.com).

The two proportions are $86/437 = 0.1968$ and $169/439 = 0.3850$. The two-sided 99% value from a standard normal distribution is 2.5758, so the approximate 99% confidence interval for the difference in proportions is 0.1882 ± 0.0773; approximately $(0.11, 0.27)$. Since this interval is well clear of 0, we can conclude that there is a real difference in this characteristic of the pots produced by the two groups of islanders.

3.5.2 Relative Risk

If the two proportions are very small, then their difference will also be small. Yet that small difference may be important. For example, knowing that the chance of survival with treatment A is 2%, whereas the chance of survival with treatment B is 1%, would strongly encourage the use of treatment A, even though the difference in the probabilities of survival is just 1%. A report of this situation would state that treatment A was twice as successful as treatment B.

In a medical context, the ratio of the success probabilities, $R = p_1/p_2$, is called the relative risk. In the notation of the 2 × 2 table, R is estimated by:

$$\widehat{R} = \frac{a/m}{c/n} = \frac{an}{cm}. \tag{3.5}$$

The distribution of $\widehat{R}$ is discrete and depends upon the values of p_1 and p_2. However, a normal approximation can be used for the distribution of $\ln(\widehat{R})$. Following a review of many alternatives, Fagerland and Newcombe (2013) found that Equation (3.6) (with z_0 denoting the appropriate critical value of the standard normal distribution) gave reliable results:

$$\exp\left(\ln(\widehat{R}) \pm 2 \sinh^{-1} \left\{ \frac{1}{2}z_0 \sqrt{\frac{1}{a} + \frac{1}{c} - \frac{1}{m+1} - \frac{1}{n+1}} \right\} \right). \tag{3.6}$$

Notice that the relative risk for failures $(1 - p_1)/(1 - p_2)$ is different to that for successes.

Example 3.5   The opinion of cancer survivors concerning the effects of stress

A study questioned cancer survivors about the factors that they believed had led to their cancer. Amongst 165 male prostate cancer survivors, 37 (22.4%) claimed that stress was a factor. Amongst 416 female breast cancer survivors, 165 (39.7%) stated that stress was a factor. The data are presented as a 2 × 2 table in Table 3.4.

TABLE 3.4 **The responses of cancer survivors who were asked whether they believed that stress had been a factor causing their cancer**

	Stress a factor	Stress not a factor	Total
Male prostate cancer	37	128	165
Female breast cancer	165	251	416
Total	202	379	581

Source: Wold et al., 2005. Reproduced with permission of Springer.

If they were correct in their beliefs then (with respect to the particular cancers) the relative risk of stress-induced cancer for women as opposed to men would be $\frac{165/416}{37/165} = 1.77$. Using Equation (3.6), with $z_0 = 1.96$, the approximate 95% confidence interval is (1.30, 2.40). Since the interval comfortably excludes 1 we can conclude that there is a distinct difference in the judgements made by males and females in this context.

3.5.3 Odds-Ratio

The *odds* on an occurrence is defined as the probability of its happening divided by the probability that it does not happen: $p/(1-p)$. For a fair coin, the probability of obtaining a head as opposed to a tail is 0.5; the odds are therefore $0.5/0.5 = 1$. Bewilderingly, in this case the odds are referred to as "evens" (and the value of the odds, 1, is an odd number)!

An *odds-ratio* θ is the ratio of the odds under one set of conditions to the odds under another set of conditions. It is given by:

$$\theta = \frac{p_1/(1-p_1)}{p_2/(1-p_2)}.$$

Yule (1900) termed it the *cross-product ratio*. By contrast with the difference in proportions (which gives different values dependent on whether we consider rows or columns) the odds-ratio is unaffected if rows and columns are interchanged. As with the relative risk, the value of the odds-ratio depends upon which category is defined as a success. If we swap categories, then the value of the odds-ratio changes from θ to $1/\theta$. This swap would not change the absolute magnitude of $\ln(\theta)$, however.

An odds-ratio of 1 corresponds to the probability of an event occurring being the same for the two conditions under comparison. Thus, when the row and column variables of a 2×2 table are independent, the odds-ratio is 1. Odds-ratios (or, rather, their logarithms) underlie the models discussed in Chapters 11–16.

In the notation of the 2 × 2 table, the obvious sample estimate of θ is

$$\frac{a/b}{c/d} = \frac{ad}{bc}.$$

However there would be a problem if any of a, b, c, or d were equal to 0, since then the estimate of the odds-ratio would be either 0 or ∞. A simple solution is to add a small amount, δ, to every cell frequency.

3.5.3.1 Gart's Method   Gart (1966) proposed using $\delta = 0.5$ to give the estimate:

$$\widehat{\theta_G} = \frac{(a + 0.5)(d + 0.5)}{(b + 0.5)(c + 0.5)}. \tag{3.7}$$

The distribution of $\widehat{\theta_G}$ is difficult to work with, but the distribution of $\ln(\widehat{\theta_G})$ is approximately normal with variance equal to the sum of the reciprocals of the entries in the 2 × 2 table. This variance would be infinite if any of the entries were zero, but the addition of 0.5 to each frequency again works well (Agresti, 1999). With z_0 denoting the appropriate critical value from a standard normal distribution, the confidence interval is

$$\exp\left(\ln(\widehat{\theta_G}) \pm z_0 \sqrt{\frac{1}{a + 0.5} + \frac{1}{b + 0.5} + \frac{1}{c + 0.5} + \frac{1}{d + 0.5}} \right). \tag{3.8}$$

Although $\widehat{\theta_G}$ has a number of desirable properties (Gart and Zweifel, 1967), Fagerland and Newcombe (2013) demonstrated that the average width of a "95% confidence interval" generated using this method was appreciably greater than 95%. They suggested the modifications given in Section 3.5.3.2.

3.5.3.2 The Fagerland–Newcombe Method   For this method we use the estimate

$$\widehat{\theta_F} = \frac{(a + 0.6)(d + 0.6)}{(b + 0.6)(c + 0.6)} \tag{3.9}$$

together with a slightly more complicated expression for the approximate confidence interval:

$$\exp\left(\ln(\widehat{\theta_F}) \pm 2\sinh^{-1}\left\{\frac{1}{2}z_0\sqrt{\frac{1}{a+0.4}+\frac{1}{b+0.4}+\frac{1}{c+0.4}+\frac{1}{d+0.4}}\right\}\right).$$

(3.10)

The interval obtained is still approximate, but, on average, will be slightly shorter. In most applications the differences between the results will be minimal.

Example 3.6   The usefulness of elastic compression stockings

A recent trial aimed to prevent post-thrombotic syndrome (PTS), which is a chronic disorder that may develop in patients after deep venous thrombosis. The trial compared the effectiveness of elastic compression stockings (ECS) with placebo stockings (i.e., standard stockings with no special treatment). Of 409 patients with ECS, 44 suffered PTS in the next 2 years. Of 394 patients with placebo stockings, 37 suffered PTS in the same period. The question of interest was whether the ECS provided any benefit. The data are set out as a 2 × 2 table in Table 3.5.

TABLE 3.5   The effectiveness of elastic stockings for the prevention of post-thrombotic syndrome (PTS)

	PTS	No PTS	Total
ECS	44	365	409
Placebo	37	357	394
Total	81	722	803

Source: Cate-Hoek, 2014. Reproduced with permission of Elsevier.

Using Equation (3.7) we obtain $\widehat{\theta_G} = 1.16$. For a 95% confidence interval $z_0 = 1.96$; so that, using Equation (3.8), we obtain the 95% confidence interval as (0.73, 1.84). Using Equation (3.9), the estimated odds-ratio is again 1.16. However, using Equation (3.10), we get the slightly narrower interval (0.74, 1.83). The difference in intervals is minimal because of the large sample sizes. Since both intervals easily include 1, we conclude that there is little evidence that the ECS are of use for reducing PTS.

REFERENCES

Agresti, A. (1999) On logit confidence intervals for the odds-ratio with small samples. *Biometrics*, **55**, 597–602.

Benjamini, Y., and Hochberg, Y. (1995) Controlling the false discovery rate: a practical and powerful approach to multiple testing. *J. R. Stat. Soc. B*, **57**, 289–300.

Blaker, H. (2000) Confidence curves and improved exact confidence intervals for discrete distributions. *Can. J. Stat.*, **28**, 783–798.

Blaker, H. (2001) Corrigenda: confidence curves and improved exact confidence intervals for discrete distributions. *Can. J. Stat.*, **29**, 681.

Cormack, R. S., and Mantel, N. (1991) Fisher's exact test: the marginal totals as seen from two different angles. *J. R. Stat. Soc., Ser. D*, **40**, 27–34.

Fagerland, M. W., and Newcombe, R. G. (2013) Confidence intervals for odds ratio and relative risk based on the inverse hyperbolic sine transformation. *Statist. Med.*, **32**, 2823–2836.

Fisher, R. A. (1934) *Statistical Methods for Research Workers*, 4th ed. (14th ed., 1970), Oliver and Boyd, Edinburgh.

Gart, J. J. (1966) Alternative analyses of contingency tables. *J. R. Stat. Soc. B*, **28**, 164–179.

Gart, J. J., and Zweifel, J. R. (1967) On the bias of various estimators of the logit and its variance. *Biometrika*, **54**, 181–187.

Irwin, J. O. (1935) Tests of significance for differences between percentages based on small numbers. *Metron*, **12**, 83–94.

Rice, W. R. (1988) A new probability model for determining exact P-values for 2×2 contingency tables when comparing binomial proportions. *Biometrics*, **44**, 1–22.

Upton, G. J. G. (1982) A comparison of tests for the 2×2 comparative trial. *J. R. Stat. Soc. A*, **45**, 86–105.

Upton, G. J. G. (1992) Fisher's exact test. *J. R. Stat. Soc. A*, **155**, 395–402.

Yates, F. (1984) Tests of significance for 2×2 contingency tables. *J. R. Stat. Soc. A*, **147**, 426–463.

Yerushalmy, J. (1947) Statistical problems in assessing methods of medical diagnosis, with special reference to X-ray techniques. *Public Health Rep.*, **62**, 1432–1449.

Youden, W. J. (1950) Index for rating diagnostic tests. *Cancer*, **3**, 32–35.

Yule, G. U. (1900) On the association of attributes in statistics. *Phil. Trans. A*, **194**, 257–319.

CHAPTER 4

THE $I \times J$ CONTINGENCY TABLE

This chapter is concerned with cases where there are two classifying variables, one with I categories and the other with J categories. The data consist of the frequencies with which the $I \times J$ possible category combinations occur. In this chapter both I and J will be greater than 2. We considered 2×2 tables in Chapter 3 and will look at $2 \times J$ tables in Chapter 7.

4.1 NOTATION

Consider a table with I rows, J columns, and IJ cells. Denote the classifying variables as A and B, and denote the categories as $A_1, A_2, \ldots, A_I$ and $B_1, B_2, \ldots, B_J$. Let the observed frequency in cell (i, j) be f_{ij} (i.e., there are f_{ij} items that simultaneously belong to categories A_i and B_j). Using the suffix 0 to denote totals gives:

$$\sum_i f_{ij} = f_{0j}, \quad \sum_j f_{ij} = f_{i0}, \quad \sum_i \sum_j f_{ij} = f_{00}.$$

The table of frequencies is assumed to be a random sample from a population for which the probability of an item being in cell (i, j) is p_{ij}, with

$$\sum_i p_{ij} = p_{0j}, \quad \sum_j p_{ij} = p_{i0}, \quad \sum_i \sum_j p_{ij} = p_{00} = 1.$$

Categorical Data Analysis by Example, First Edition. Graham J. G. Upton.
© 2017 John Wiley & Sons, Inc. Published 2017 by John Wiley & Sons, Inc.

4.2 INDEPENDENCE IN THE $I \times J$ CONTINGENCY TABLE

4.2.1 Estimation and Degrees of Freedom

The probability of an item belonging to category i of variable A is p_{i0}. Similarly, the probability of an item belonging to category j of variable B is p_{0j}. Now, if two events are independent, then the probability that both occur is the product of their probabilities. Therefore, under independence,

$$p_{ij} = p_{i0}p_{0j}. \tag{4.1}$$

The obvious estimates of p_{i0} and p_{0j} are $\widehat{p_{i0}} = f_{i0}/f_{00}$ and $\widehat{p_{0j}} = f_{0j}/f_{00}$, respectively. Using these probability estimates, under independence, the probability of the simultaneous occurrence of category i of variable A and category j of variable B would be

$$\frac{f_{i0}}{f_{00}} \times \frac{f_{0j}}{f_{00}} = \frac{f_{i0}f_{0j}}{f_{00}^2}.$$

Since there are f_{00} observations in total it follows that, if we know that the classifying variables are independent, and we know the row and column totals, then our best guess of the frequency in cell (i,j) would be

$$e_{ij} = \frac{f_{i0}f_{0j}}{f_{00}^2} \times f_{00} = \frac{f_{i0}f_{0j}}{f_{00}}. \tag{4.2}$$

Consider the row sum

$$e_{i0} = e_{i1} + e_{i2} + \cdots + e_{iJ}$$

and substitute using Equation (4.2). The result is

$$\begin{aligned}
e_{i0} &= \frac{f_{i0}f_{01}}{f_{00}} + \frac{f_{i0}f_{02}}{f_{00}} + \cdots + \frac{f_{i0}f_{0J}}{f_{00}} \\
&= \frac{f_{i0}}{f_{00}}(f_{01} + f_{02} + \cdots + f_{0J}) \\
&= \frac{f_{i0}}{f_{00}} \times f_{00} \\
&= f_{i0}.
\end{aligned}$$

This is true for any value of i. In the same way we find that $e_{0j} = f_{0j}$ for any value of j. Thus, for the independence model, *the observed marginal totals are exactly reproduced by the marginal totals of the expected frequencies*. This is an example of the general result discussed in Section 13.1.

The IJ expected frequencies are therefore subject to the constraints that they must sum to prescribed totals. There are I row constraints and J column constraints, which makes it appear that there are $(I + J)$ constraints in all. However, we can "save" one constraint, since, once we know all the row totals, we know the value of f_{00}, and, once we know $(J - 1)$ of the column totals, we can deduce the value of the last column total. In general, for an $I \times J$ table, there are therefore $I + J - 1$ constraints. With IJ cell frequencies and $I + J - 1$ constraints there are therefore

$$IJ - (I + J - 1) = (I - 1)(J - 1)$$

degrees of freedom for the expected values under the model of independence.

4.2.2 Odds-Ratios and Independence

Consider any two rows (i and i', say) and any two columns (j and j', say). There are four cells at the intersections of these rows and columns: (i,j), (i,j'), (i',j), and (i',j'). The odds-ratio for the probabilities of these four combinations is

$$p_{ij}p_{i'j'}/p_{ij'}p_{i'j}.$$

Substitution using Equation (4.1) gives

$$\{(p_{i0}p_{0j})(p_{i'0}p_{0j'})\}/\{(p_{i'0}p_{0j})(p_{i0}p_{0j'})\} = 1.$$

Thus, under independence, *every* odds-ratio is equal to 1.

4.2.3 Goodness of Fit and Lack of Fit of the Independence Model

4.2.3.1 Tests of Goodness-of-Fit   Substitution of the expected frequencies given by Equation (4.2) into the general formula for Pearson's X^2 statistic given by Equation (2.1) gives

$$X^2 = \sum_i \sum_j \left\{ \left(f_{ij} - \frac{f_{i0}f_{0j}}{f_{00}} \right)^2 \bigg/ \frac{f_{i0}f_{0j}}{f_{00}} \right\} = \frac{1}{f_{00}} \sum_i \sum_j \frac{(f_{ij}f_{00} - f_{i0}f_{0j})^2}{f_{i0}f_{0j}}.$$

$$(4.3)$$

In this case, since there are $(I + J - 1)$ constraints and each constraint reduces the number of degrees of freedom by one (see Section 2.1.2), the distribution of X^2 is approximately χ^2-squared with $(I - 1)(J - 1)$ degrees of freedom.

In the same way, for the likelihood-ratio goodness-of-fit statistic, we have:

$$G^2 = 2 \sum_i \sum_j f_{ij} \ln \left(f_{ij} \bigg/ \frac{f_{i0}f_{0j}}{f_{00}} \right),$$ (4.4)

with an approximate χ^2-squared distribution with $(I - 1)(J - 1)$ degrees of freedom. This formula can be usefully rewritten by expanding the logarithm to give

$$G^2 = 2 \sum_i \sum_j f_{ij} \ln(f_{ij}) + 2 \sum_i \sum_j f_{ij} \ln(f_{00}) - 2 \sum_i \sum_j f_{ij} \ln(f_{i0})$$
$$- 2 \sum_i \sum_j f_{ij} \ln(f_{0j}).$$

Simplifying the summations gives

$$G^2 = 2 \sum_i \sum_j f_{ij} \ln(f_{ij}) + 2f_{00} \ln(f_{00}) - 2 \sum_i f_{i0} \ln(f_{i0}) - 2 \sum_j f_{0j} \ln(f_{0j}). \quad (4.5)$$

4.2.3.2 Residuals

Regardless of whether or not a model appears to fit the data, we should always compare every observed cell frequency with the corresponding frequency estimated from the model. In this case that implies comparing f_{ij} with $e_{ij} = f_{i0}f_{0j}/f_{00}$. The size of a residual, $(f_{ij} - e_{ij})$, is partly a reflection of the number of observations: the difference between 5 and 15 is the same as the difference between 1005 and 1015, but the latter is a trivial proportionate error. More relevant is:

$$r_{ij} = \frac{f_{ij} - e_{ij}}{\sqrt{e_{ij}}}.$$ (4.6)

Since Pearson's goodness-of-fit statistic X^2 is equal to $\sum \sum r_{ij}^2$, r_{ij} is described as a *Pearson residual*. If the value of e_{ij} had been independent of the data, then r_{ij} would be (approximately) an observation from a standard normal distribution. However, to take account of the fact that e_{ij} is a function of the observed data, a correction is required. The result is called a *standardized residual* and is given by:

$$s_{ij} = \frac{f_{ij} - e_{ij}}{\sqrt{e_{ij} \left(1 - \frac{f_{i0}}{f_{00}} \right) \left(1 - \frac{f_{0j}}{f_{00}} \right)}}.$$ (4.7)

If the variables are truly independent, then approximately 95% of the s_{ij}-values will lie in the range $(-2, 2)$. A value with magnitude greater than 3 only rarely occurs by chance, while a value with magnitude greater than 4 occurs by chance on less than one occasion in a million.

Example 4.1   Political affiliation and newspaper choice

Table 4.1 examines the relationship between an individual's choice of daily newspaper and their political party affinity.

TABLE 4.1   Cross-tabulation of political party affinity and daily newspaper readership in the United Kingdom in 2003–2004

Political party affinity	Daily newspaper readership				
	Broadsheet	Middle-market	Tabloid	Local paper	Total
Conservative	25	47	12	10	94
Labour	41	17	60	24	142
Liberal Democrat	13	4	3	6	26
None	73	69	117	62	321
Total	152	137	192	102	583

Source: SN 5073, Survey of Public Attitudes towards Conduct in Public Life, 2003–2004. Reproduced with permission of the UK Data Service.
Broadsheet papers include *The Daily Telegraph*, *The Guardian*, *The Independent*, and *The Times*. Middle-market papers include the *Daily Mail* and *Daily Express*. Tabloids include *The Sun* and *Daily Mirror*.

Under the hypothesis of independence, the expected number of Conservative broadsheet readers in this sample would be $94 \times 152/583 = 24.51$, which is very close to the 25 actually observed. However, the number of Conservative tabloid readers would be $94 \times 192/583 = 30.96$ which is much greater than the 12 observed. The goodness-of-fit statistics have values 68.7 (for X^2) and 67.3 (for G^2). For both statistics the reference chi-squared distribution has $(4 - 1)(4 - 1) = 9$ degrees of freedom. These observed values correspond to tail probabilities of the order of 10^{-11} so there is no doubt that newspaper readership and political affinity are not independent (though whether it is the newspaper choice that shapes a reader's political views or whether it is the political views that govern newspaper choice is not something that we can discover with these data).

Table 4.2 gives the full set of expected frequencies and also the standardized residuals obtained using Equation (4.7). Standardized residuals greater than 3 in magnitude (which would appear by chance on about 3 times in a thousand) are shown in bold. It appears that the major cause of the lack of independence is the fact that the number of Conservative supporters reading middle-market newspapers is about twice that expected under independence.

TABLE 4.2   Expected frequencies under independence and the corresponding standardized residuals for the data of Table 4.1

Expected frequencies				Standardized residuals			
24.51	22.09	30.96	16.45	0.13	**6.62**	−4.54	−1.91
37.02	33.37	46.77	24.84	0.87	**−3.73**	2.72	−0.21
6.78	6.11	8.56	4.55	2.84	−1.00	−2.37	0.77
83.69	75.43	105.72	56.16	−2.03	−1.26	2.00	1.28

4.3 PARTITIONING

If U and V are independent random variables, with U having a χ_u^2-distribution and V having a χ_v^2-distribution, then their sum $U + V$ has a $\chi_{(u+v)}^2$-distribution. The implication is that, since the goodness-of-fit statistics X^2 and G^2 have approximate χ^2-distributions, we can potentially identify the sources of any lack of fit by partitioning a table into independent subtables. Thus, for a 2×3 table

$$
\begin{array}{ccc|c}
f_{11} & f_{12} & f_{13} & f_{10} \\
f_{21} & f_{22} & f_{23} & f_{20} \\
f_{01} & f_{02} & f_{03} & f_{00}
\end{array}
\equiv
\begin{array}{cc|c}
f_{11} & f_{12}+f_{13} & f_{10} \\
f_{21} & f_{22}+f_{23} & f_{20} \\
f_{01} & f_{02}+f_{03} & f_{00}
\end{array}
+
\begin{array}{cc|c}
f_{12} & f_{13} & f_{12}+f_{13} \\
f_{22} & f_{23} & f_{22}+f_{23} \\
f_{02} & f_{03} & f_{02}+f_{03}
\end{array} .
$$

Between them, the two right-hand tables contain all the information presented in the left-hand table. Importantly, they contain no other information. Both presentations are equally valid and should therefore lead to corresponding conclusions. For the 2×3 table there are $1 \times 2 = 2$ degrees of freedom. Each 2×2 table has a single degree of freedom: as usual $1 + 1 = 2$.

For each of the three tables, the values of X^2 and G^2 will be probably very similar. However, it is here that G^2 has an advantage over X^2, since the values of G^2 for the two 2×2 tables will always *exactly* equal that for the 2×3 table, whereas that is not the case for the values of X^2.

4.3.1 *Additivity of G^2

Applying Equation (4.5) to the 2×3 table gives

$$
2\{f_{11}\ln(f_{11}) + f_{12}\ln(f_{12}) + f_{13}\ln(f_{13}) + f_{21}\ln(f_{21}) + f_{22}\ln(f_{22}) + f_{23}\ln(f_{23})
$$

$$
+ f_{00}\ln(f_{00}) - f_{10}\ln(f_{10}) - f_{20}\ln(f_{20}) - f_{01}\ln(f_{01}) - f_{02}\ln(f_{02}) - f_{03}\ln(f_{03})\}.
$$

Applying Equation (4.5) to the first 2×2 table gives

$$2\{f_{11} \ln(f_{11}) + (f_{12} + f_{13}) \ln(f_{12} + f_{13} + f_{21} \ln(f_{21} + (f_{22} + f_{23}) \ln(f_{22} + f_{23})$$
$$+ f_{00} \ln(f_{00}) - f_{10} \ln(f_{10}) - f_{20} \ln(f_{20}) - f_{01} \ln(f_{01}) - (f_{02} + f_{03}) \ln(f_{02} + f_{03})\},$$

and to the second 2×2 table gives

$$2\{f_{12} \ln(f_{12}) + f_{13} \ln(f_{13}) + f_{22} \ln(f_{22}) + f_{23} \ln(f_{23}) + (f_{02} + f_{03}) \ln(f_{02} + f_{03})$$
$$- (f_{12} + f_{13}) \ln(f_{12} + f_{13}) - (f_{22} + f_{23}) \ln(f_{22} + f_{23}) - f_{02} \ln(f_{02}) - f_{03} \ln(f_{03})\}.$$

Thus the contribution of $+(f_{22} + f_{23}) \ln(f_{22} + f_{23})$ to G^2 for the first 2×2 table is cancelled out by the contribution of $-(f_{22} + f_{23}) \ln(f_{22} + f_{23})$ for the second 2×2 table. It is this, and the corresponding cancellations for the other introduced summations that result in precise equality of the G^2-values for the complete table and for its components.

Example 4.1   Political affiliation and newspaper choice (*continued*)

The lack of independence between political affiliation and newspaper readership in Table 4.1 was very apparent and easy to identify. The mean of a χ^2 distribution with d degrees of freedom is d, and its standard deviation is $\sqrt{d}$, so our observed values of 68.7 for X^2 and 67.3 for G^2 were much greater than would have been expected by chance. We might wonder just how much of that lack of fit was due to the apparent Conservative preference for middle-market newspaper. The answer is provided by splitting the original table into parts (partitioning), with one part focusing on the (Conservative, middle-market newspaper) combination.

The partitioned data appear in Table 4.3. A consequence of the partitioning is the construction of omnibus categories marked "Other." The totals for these categories are shown emboldened. Notice that each of these totals appears not only in the body of one subtable, but also in the margin of another subtable. It is this feature that ensures that the component G^2-values sum correctly to the G^2-value for the complete table.

The corresponding goodness-of-fit statistics are summarized in Table 4.4. The first four rows of the table show the goodness-of-fit values for the independence model applied separately to each subtable. Included in the table is the ratio of G^2 divided by the degrees of freedom. This ratio is a useful guide since the value should be near 1 if the model is a good fit. For subtables B, C, and D, the ratios are all between 3 and 5 (corresponding to tail probabilities

TABLE 4.3 Partitioned version of Table 4.1 using four sub-tables. The emboldened values did not appear in Table 4.1

A	Newspaper			B	Newspaper			
Party	Middle-market	Other	Total		Broad-sheet	Tabloid	Local	Total
Con.	47	**47**	94		25	12	10	**47**
Other	**90**	**399**	**489**		**127**	**180**	**92**	**399**
Total	137	**446**	583		52	192	102	**446**

C	Newspaper			D	Newspaper			
Party	Middle-market	Other	Total		Broad-sheet	Tabloid	Local	Total
Lab.	17	**125**	142		41	60	24	**125**
Lib. Dem.	4	**22**	26		13	3	6	**22**
None	69	**252**	32		73	117	62	**252**
Total	**90**	**399**	**489**		**127**	**180**	**92**	**399**

ranging between 1% and 4%) and indicating a modest lack of independence. For subtable A, however, the ratio is nearly 40 (corresponding to a tail probability of the order of 10^{-10}); this confirms that the preference of Conservative supporters for middle-market newspapers is the dominant feature.

The final rows in Table 4.4 report the sum of the goodness-of-fit values of the separate tables with that for the original table. The apparent difference of 0.0001 in the values for G^2 is the result of summing the results rounded to four decimal places; the true difference is zero. By contrast there is a marked difference in the sum of the X^2 values.

TABLE 4.4 Goodness-of-fit statistics for the independence model applied to the four subtables of Table 4.1 given in Table 4.3

Subtable	d.f.	G^2	$G^2/\text{d.f.}$	X^2
A	1	38.4617	38.5	43.7813
B	2	9.2774	4.6	9.4332
C	2	6.4753	3.3	6.1126
D	4	13.0603	3.3	12.2185
A+B+C+D	9	67.2747		71.5456
Table 4.1	9	67.2748		68.7099

4.3.2 Rules for Partitioning

There is really only one rule:

Every cell in the original table must appear exactly once as a cell in a subtable.

A check on this is provided by the degrees of freedom of the subtables correctly summing to the degrees of freedom of the original table. Table 4.4 demonstrated this. A critical feature of the subtables is that any cell frequency that appears in a subtable without appearing in the original table will also be found as a marginal total for another subtable (see Table 4.3 for an example).

4.4 GRAPHICAL DISPLAYS

Diagrams are often useful aids for interpreting data. In this section two possibilities are presented.

Mosaic plots give the viewer an idea of the respective magnitudes of the categories of the classifying variables, together with an idea of which specific category combinations are unusually scarce or frequent.

Cobweb diagrams focus on identifying the unusually scarce or frequent category combinations.

4.4.1 Mosaic Plots

The *mosaic plot* was introduced by Friendly (1994) in order to give a visual impression of the relative sizes of the $I \times J$ frequencies. It also gives an indication of how the observed table differs from a table displaying independence. The precise details depend on the statistical package being used. Using the `mosaicplot` function in R, positive and negative residuals are distinguished by the use of continuous or dotted borders to the mosaic boxes, with shading indicating the magnitude of a residual.

Figure 4.1 shows the mosaic plot for the data of Table 4.1. It was produced by the following R commands:

```
R code
Newspaper<-c("Broadsheet","Middle-market","Tabloid","Local
    paper");
Party<-c("Conservative","Labour","Liberal Democrat",
    "None");
Freq<-c(25,47,12,10,41,17,60,24,13,4,3,6,73,69,117,62);
```

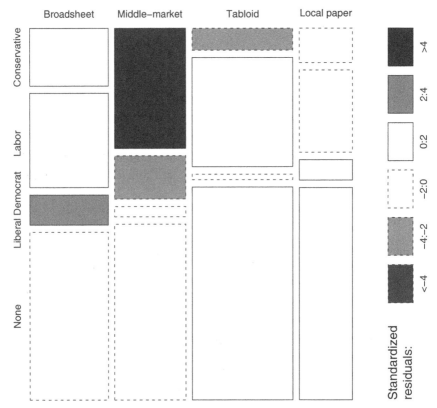

FIGURE 4.1 Mosaic diagram, using standardized residuals (Equation 4.7) for the independence model applied to the data of Table 4.1.

```
t<-array(Freq,dim=c(4,4),dimnames=list(Newspaper,Party));
pdf("mosaic.pdf", colormodel="gray");
mosaicplot(t,main="",type="pearson",shade=TRUE,color=FALSE);
dev.off();
```

Friendly (2000) extended the mosaic approach to tables with more than two variables.

4.4.2 Cobweb Diagrams

The cobweb diagram, which focuses on the interactions between pairs of variables, was introduced by Upton (2000). In the diagram, lines indicate category combinations having standardized residuals (Equation 4.7) with magnitudes greater than 2. The thickness of the line is governed by the magnitude of the residual, with black lines indicating unusually common combinations and gray lines indicating unusually scarce combinations. The R code for the cobweb function is provided in the Appendix.

Example 4.1   Political affiliation and newspaper choice (*continued*)

Figure 4.2 illustrates the relationships between the variable categories. The code used to apply the function in this instance is:

R code (*continued*)

```
df<-data.frame(expand.grid(Newspaper=Newspaper,Party=Party),
    Freq,stringsAsFactors = TRUE);
scale<-1;
cobweb(df,scale,"paper.ps")
```

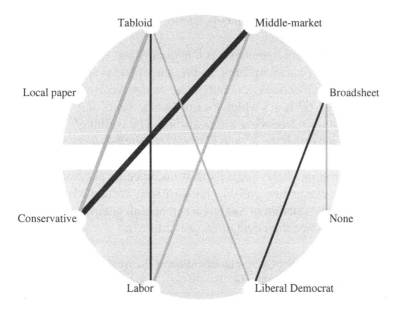

Newspaper

Party

FIGURE 4.2   Cobweb diagram for the independence model applied to the data of Table 4.1. The lines indicate the category combinations that have the largest standardized residuals (Black = positive, Gray = negative).

4.5 TESTING INDEPENDENCE WITH ORDINAL VARIABLES

Table 4.5 reports the result of a cross-classification of objects by their color and type.

TABLE 4.5   Cross-classification of the colors and types of 78 objects

	Color			
	Red	White	Blue	Green
Type A	2	4	5	9
Type B	3	5	7	9
Type C	7	5	4	3
Type D	8	4	2	1

We can test the null hypothesis that color and type are independent of one another by using the X^2 goodness-of-fit test (Section 4.2.3.1). In this case $X^2 = 16.6$. With $(4 - 1)(4 - 1) = 9$ degrees of freedom, we find that 16.6 is not an exceptionally large value (a tail probability of about 5.6%): we conclude that the null hypothesis of independence can be accepted. This seems a not unreasonable conclusion.

For Table 4.6, the null hypothesis is that quality and expense are independent of one another. This seems implausible, since there is an obvious connection: better-quality objects are generally more expensive. However, the counts are the same as for Table 4.5 which means that the value of X^2 is again 16.6. Our standard test of independence appears not to be working!

The explanation is that the X^2 and G^2 tests are *omnibus tests*: they are comparing a specific situation (independence) against *all possible* alternatives. That is not appropriate here, since there is one specific alternative that is of interest (namely, that quality and cost are positively correlated). The intrinsic difference from the situation of Table 4.5 is that both quality and expense are ordinal variables (since their categories are ordered).

TABLE 4.6   Cross-classification of the quality and expense of 78 objects

	Quality of manufacture			
	Very low	Low	High	Very high
Very expensive	2	4	5	9
Expensive	3	5	7	9
Cheap	7	5	4	3
Very cheap	8	4	2	1

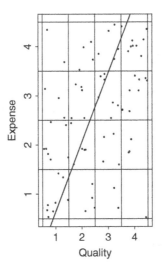

FIGURE 4.3 A scatter diagram representation of the data in Table 4.6, with a line suggesting the dependence between the two variables.

Suppose, now, that quality and expense were really continuous variables with values that had been recorded (1, 2, 3, and 4, say) with limited accuracy. Thus a value recorded as "Very expensive" was measured as some value between 3.5 and 4.5. The entire data set might have looked as shown in Figure 4.3, with two data points in the top left cell, one data point in the bottom right, and so on (matching the counts in Table 4.6).

Using the data illustrated in the scatter diagram, a natural measure of the relation between quality and expense would be the coefficient of correlation r. We can apply this idea to the data summarized in Table 4.6. Denote the classifying variables as X and Y and ascribe the value i to the ith category of X and the value j to the jth category of Y. With the count in cell (i,j) being f_{ij}, the squared correlation is then given by

$$r^2 = \frac{\left(f_{00}\sum_i\sum_j ijf_{ij} - \left(\sum_i if_{i0}\right)\left(\sum_j jf_{0j}\right)\right)^2}{\left(f_{00}\sum_i i^2f_{i0} - \left(\sum_i if_{i0}\right)^2\right)\left(f_{00}\sum_j j^2f_{0j} - \left(\sum_j jf_{0j}\right)^2\right)}. \tag{4.8}$$

A test of the hypothesis that $r = 0$ (corresponding to independence between the variables) against the alternative $r \neq 0$ is then provided by

$$M^2 = (f_{00} - 1)r^2, \tag{4.9}$$

which has an approximate χ_1^2-distribution. The 1 degree of freedom here can be thought of as being a specific 1 of the $(I-1)(J-1)$ degrees of freedom for the general goodness-of-fit test: effectively we are partitioning the latter into a component (M^2) that encapsulates an underlying linear (or, at least, monotonic) relation between the classifying variables, and the balance referring to remaining deviations from independence.

For the data in Table 4.6, $r^2 = 0.19$, giving $M^2 = 14.6$ which corresponds to a tail probability of about 0.0001. There is therefore very clear evidence of a monotonic association between the variables with higher quality associated with higher expense and lower quality with lower expense.

REFERENCES

Friendly, M. (1994) Mosaic displays for multi-way contingency tables. *J. Am. Stat. Ass.*, **89**, 190–200.

Friendly, M. (2000) *Visualizing Categorical Data*, SAS Institute Inc., Cary, NC.

Upton, G. J. G. (2000) Cobweb diagrams for multi-way contingency tables. *J. R. Stat. Soc. D*, **49**(1), 79–85.

CHAPTER 5

THE EXPONENTIAL FAMILY

This chapter gives a very brief introduction to the exponential family. The existence of the family is important because it includes the most relevant distributions (binomial, multinomial, and Poisson) amongst its members and because it is possible to devise standardized algorithms for fitting models to data from any distribution in the family. In this chapter, only the briefest sketch is given of these algorithms since the associated theory is not otherwise relevant to the contents of the book. Readers uninterested in this background theory should move hastily to the next chapter!

5.1 INTRODUCTION

The most familiar linear model is the *linear regression* model:

$$\mathrm{E}(Y) = \alpha + \beta x.$$

Here a straight line describes the manner in which the expectation of the *response variable Y* depends on the value of an *explanatory variable x*. The parameters α (the intercept) and β (the slope) are unknown and require estimation using the pairs of observations (x_1, y_1), (x_2, y_2), ..., (x_n, y_n). In the commonest case it is assumed that the observation pairs are independently sampled, with each value of Y arising from a normal (Gaussian) distribution

Categorical Data Analysis by Example, First Edition. Graham J. G. Upton.
© 2017 John Wiley & Sons, Inc. Published 2017 by John Wiley & Sons, Inc.

with variance σ^2. Typically, the values of x are themselves values from a continuous variable. Thus Y might refer to the weight of a person and x to that person's height.

The extension to the case of several continuous explanatory variables is referred to as *multiple regression*. For example:

$$E(Y) = \alpha + \beta_1 x_1 + \beta_2 x_2.$$

If the response variable has a normal distribution, but all the explanatory variables are categorical (as for example when Y refers to the volume of a tomato crop, x_1 refers to the variety of tomato, and x_2 to the type of fertilizer used), then the resulting linear model is described as an *analysis of variance model (ANOVA)*. This type of situation is common with designed experiments. If the explanatory variables include a mixture of continuous and categorical variables then the resulting model may be described as an *analysis of covariance model (ANOCOVA)*.

While this book is concerned with models of these general types, there is an important difference, since the response variable is not continuous but categorical. However, since the different distributions involved are all members of the *exponential family*, the linear models used can all be analyzed using the same algorithm (Nelder and Wedderburn, 1972).

5.2 THE EXPONENTIAL FAMILY

All distributions that are members of the family have probability density functions (continuous case), or probability distributions (discrete case), that can be written in the form

$$\exp\left[a(y)b(\theta) + c(\theta) + d(y)\right], \tag{5.1}$$

where y is an observation on the random variable Y, θ is a parameter of the distribution, a and d are functions of y that do not involve θ, and b and c are functions of θ that do not involve y. In the case where $a(y) = y$, the function $b(\theta)$ is described as the *natural parameter* of the distribution; models will focus on this function.

Example 5.1   The binomial distribution

This distribution has a single parameter (the probability of occurrence of some event) denoted by p. We now write the distribution as

$$P(y) = \exp\left[y\ln\left(\frac{p}{1-p}\right) + n\ln(1-p) + \ln\binom{n}{r}\right]. \tag{5.2}$$

Thus $a(y) = y$, $b(p) = \ln\left(\frac{p}{1-p}\right)$, $c(p) = n\ln(1-p)$, and $d(y) = \ln\binom{n}{r}$. Since $a(y)$ is equal to y, the natural parameter is $\ln\left(\frac{p}{1-p}\right)$, which is called the *logit*.

Example 5.2   The Poisson distribution

This distribution has a single parameter (the mean) denoted by μ. We now write the distribution as

$$P(y) = \exp\left[y\ln(\mu) - \mu - \ln(y!)\right]. \tag{5.3}$$

Thus $a(y) = y$, $b(\mu) = \ln(\mu)$, $c(\mu) = -\mu$, and $d(y) = -\ln(y!)$. Since $a(y) = y$, $\ln(\mu)$ is the natural parameter.

5.2.1 The Exponential Dispersion Family

This is a simple extension of the exponential family resulting from the introduction of a second parameter ϕ. This wider family has probability density function (or probability distribution) with the form:

$$\exp\left[\frac{yb(\theta) + c(\theta)}{f(\phi)} + d(y; \phi)\right], \tag{5.4}$$

where f is a function, and ϕ is called the *dispersion parameter* of the distribution (Jorgensen, 1987). When the value of ϕ is known, then, with suitable function definitions, the formulation becomes that of the exponential family given by Equation (5.1).

Example 5.3   The normal distribution

Here the parameter of interest is μ with the dispersion parameter being σ. The probability density function can be written as

$$\exp\left[\frac{1}{\sigma^2}\left(y\mu - \frac{1}{2}\mu^2\right) - \left(\frac{y^2}{2\sigma^2} + \ln(\sigma\sqrt{2\pi})\right)\right],$$

with $b(\mu) = \mu$, $c(\mu) = -\frac{1}{2}\mu^2$, and $f(\sigma) = \sigma^2$.

5.3 COMPONENTS OF A GENERAL LINEAR MODEL

There are three components:

1. A random variable Y having a distribution that is a member of the exponential family.

2. A *linear predictor* which is simply a linear combination of explanatory variables (e.g., $\alpha + \beta x$).
3. A *link function*. This is a function of the parameter of interest (p for the binomial distribution or μ for the Poisson and normal distributions). It is the value of this function that is explained by the linear predictor. In a case where the link function is the natural parameter of the distribution, it is called the *canonical link function*.

Example 5.4   The Poisson distribution

Suppose that $Y_1, Y_2, \ldots, Y_I$ are Poisson random variables with means that are linear functions of J explanatory X-variables. For a Poisson distribution with mean μ the canonical link function is $\ln(\mu)$, (see Example 5.1) so, with I sets of observations, the model is

$$\ln(\mu_i) = \sum_{j=1}^{J} \beta_j x_{ji}, \qquad i = 1, 2, \ldots, I, \qquad (5.5)$$

where $\beta_1, \beta_2, \ldots, \beta_J$ are parameters that must be estimated. Models of this type are called *log-linear models* and are introduced in Chapter 11.

Example 5.5   The binomial distribution

Suppose that $Y_1, Y_2, \ldots, Y_I$ are binomial random variables with success probabilities that are linear functions of J explanatory X-variables. For a binomial distribution the canonical link function is the logit (see Example 5.2) so, with I sets of observations, the model is

$$\ln\left(\frac{p_i}{1 - p_i}\right) = \sum_{j=1}^{J} \beta_j x_{ji}, \qquad i = 1, 2, \ldots, I, \qquad (5.6)$$

where $\beta_1, \beta_2, \ldots, \beta_J$ are parameters that must be estimated. Models of this type are called *logistic regression models* or, simply, *logit models*. They are discussed in Chapters 7, 8, and 9.

5.4 ESTIMATION

Fortunately for the user, the precise details of the estimation process are not required by the analyst armed with a computer! What follows is the barest outline of the algorithm used.

The joint likelihood of n y-values from distributions that are members of the exponential family is

$$\prod_{i=1}^{n} \{\exp[a(y_i)b(\theta_i) + c(\theta_i) + d(y_i)]\},$$

so that the log-likelihood L is given by

$$L = \sum_{i=1}^{n} [a(y_i)b(\theta_i) + c(\theta_i) + d(y_i)].$$

For both the Poisson distribution and the binomial distribution, using Equation (5.5) or (5.6), $b(\theta_i) = \sum_{j=1}^{J} \beta_j x_{ji}$, so that the maximum likelihood estimates of the β-parameters are the joint solutions of

$$\frac{\partial L}{\partial \beta_1} = 0, \frac{\partial L}{\partial \beta_2} = 0, \dots , \frac{\partial L}{\partial \beta_J} = 0.$$

Finding the maximum of a function (in this case the likelihood) is a standard numerical problem. In the context of linear models the most common methods of solution are the Newton–Raphson and Fisher scoring methods. These procedures are very similar to one another and both are iterative; here is an outline for the case of a single parameter, β:

1. Begin with an educated guess, $\beta^{(0)}$, of the position of the maximum.
2. Fit a quadratic in the vicinity of $\beta^{(0)}$.
3. Determine the location of the maximum of this quadratic. Call this value $\beta^{(1)}$.
4. Fit a quadratic in the vicinity of $\beta^{(1)}$.
5. Etc.

Whichever method is used, there will usually be speedy convergence to $\hat{\beta}$, the parameter value that maximizes L. When using a standard computer package, the numerical calculations will usually be invisible to the analyst (except, perhaps, for a report of the number of iterations required for convergence).

REFERENCES

Jorgensen, B. (1987) Exponential dispersion models. *J. R. Stat. Soc. B*, **49**, 127–162.

Nelder, J. A., and Wedderburn, R. W. M. (1972) Generalized linear models. *J. R. Stat. Soc. A*, **135**, 370–384.

CHAPTER 6

A MODEL TAXONOMY

Chapter 5 focused on the exponential family and introduced the general linear model. Linear models of one sort or another underlie the remainder of this book and it is easy to lose track of which type of model (and hence which chapter of the book) is relevant for the analysis of a particular set of data. The aim of this short chapter is to help with this fundamental problem; presenting the material as a separate chapter will, it is hoped, make it easier to locate the appropriate model type.

6.1 UNDERLYING QUESTIONS

6.1.1 Which Variables Are of Interest?

Sometimes there will be an easy answer to this question: "All of them!". However, when the data form part of a large-scale survey, it can be difficult to decide where to focus one's efforts and which are the relevant variables. Since the number and type of the variables have a bearing on which model, or which book chapter, is relevant, it is a necessary first step.

6.1.2 What Categories Should Be Used?

For some variables the categories are obvious and well-defined. For example, the variable Sex can be regarded as having just two categories. However,

Categorical Data Analysis by Example, First Edition. Graham J. G. Upton.
© 2017 John Wiley & Sons, Inc. Published 2017 by John Wiley & Sons, Inc.

for most variables the situation is more difficult. Consider, for example, the question "Do you regard yourself as belonging to any particular religion?" The annual British Social Attitudes Survey provides for 18 possible answers in addition to the usual "Refusal" and "Don't know" replies. With several thousand respondents for each year's survey, there are nevertheless several categories where the number of respondents is in single figures. Perhaps these categories should be combined into a single "Other" category? That is a decision that will depend on the context and purpose of the data analysis.

Decisions concerning grouping categories also affect variables that are inherently continuous (e.g., Age). If a survey of a human population records age in years, then there will certainly be a need for grouping into age bands (e.g., 25–34): a good choice should ensure that (if possible, and subject to the purpose of the data collection) there are large numbers in each group, so that the model parameters can be estimated with confidence.

6.1.3 What Is the Type of Each Variable?

In the context of categorical data there are two types of variable: *nominal* (names such as Red, White, and Blue) and *ordinal* (such as Small, Medium, and Large).

6.1.4 What Is the Nature of Each Variable?

Is a particular variable (*A*, say) included because we want to understand what affects the choice of category for that variable, or is the variable (*B*, say) included because it may influence the choice of category for *A*? Variable *A* might be referred to as a *dependent variable*, or *response variable*; variable *B* might be called an *explanatory variable* or *factor*.

Sometimes a variable may be a link in a chain: the category of *B* might be influenced by a third variable *C*. In such a case variable *B* is both a factor (affecting *A*) and a response (affected by *B*).

TABLE 6.1 **A guide to the alternative model types for categorical response variables**

Number of explanatory variables	Number of response variables	Number of response categories	Relevant distribution	Type of model	Chapters
1	1	2	Binomial	Logistic regression	7, 9
> 1	1	2	Binomial	Logistic regression	8, 9
Any	1	> 2	Multinomial	Logistic regression	10
Any	> 1	Any	Poisson	Log-linear	11–16

6.2 IDENTIFYING THE TYPE OF MODEL

The choice of model type depends upon the nature of the variables involved and the numbers of categories that each variable has. As Table 6.1 indicates, if there is a single response variable, then some form of logistic regression model will be appropriate. In cases where there is more than one response variable, a log-linear model will be required.

CHAPTER 7

THE 2 × J CONTINGENCY TABLE

This chapter is concerned with the case where an explanatory variable with J categories may provide information about a two-category response variable. For the opposite situation where an explanatory variable with two categories provides information about a J-category response variable, see Chapter 10.

7.1 A PROBLEM WITH X^2 (AND G^2)

We begin with a reminder of the shortcomings (discussed in Section 4.5) of the omnibus goodness-of-fit statistics X^2 and G^2 when used with variables that have ordered categories. As an example, consider the following data:

	B_1	B_2	B_3	B_4	B_5
A_1	40	15	26	36	21
A_2	10	10	14	12	9

For these data the value of X^2 is 4.23. The reference distribution is χ_4^2, for which the upper 5% point is 9.49. At the 5% significance level, the hypothesis of independence between A and B would therefore not be rejected.

Categorical Data Analysis by Example, First Edition. Graham J. G. Upton.
© 2017 John Wiley & Sons, Inc. Published 2017 by John Wiley & Sons, Inc.

Now consider the following table:

	B_1	B_2	B_3	B_4	B_5
A_1	15	26	21	36	40
A_2	10	14	9	12	10

This is the same as the previous table, but the columns have been reordered. Since the same numbers are involved, X^2 is again equal to 4.23 and the conclusion is unaltered.

Suppose now that we are told that A refers to the rooting of cuttings, with A_1 representing success and A_2 representing failure, while B refers to the amounts of hormone rooting powder (B_1: 1 mg per cutting,..., B_5: 5 mg per cutting). It is natural to examine the success rates. There were 25 plants that received 1 mg of the powder. Of these, 15 were successes: a 60% rate. Here are the results of the previous table now expressed in terms of success rates:

	x: Amount of rooting powder (mg)				
	1	2	3	4	5
Sample size (n)	25	40	30	48	50
Success rate (p)	60%	65%	70%	75%	80%

This table conveys the same information as its predecessor, but now it is immediately apparent that there *is* an association between the variables. The Pearson test did not uncover the relation because it was unable to use the information that the categories were ordered. Of course, the same would apply to the likelihood-ratio statistic G^2 or indeed, any other test procedure that ignored the ordering of the categories.

7.2 USING THE LOGIT

In the previous example, for the reordered table the relation was particularly clear. The model

$$p = \mu + \alpha x$$

would provide a perfect fit (with $\mu = 55\%$ and $\alpha = 5\%$). We could confidently predict that 2.5 mg of powder would give a 67.5% success rate, and it seems plausible that 6 mg would give an 85% success rate. But, and this is a big BUT, what would happen with 10 mg of powder? Our model would predict $(55 + 5 \times 10 = 105)\%$!

Of course, extrapolation can always give foolish results, but we should be using a model that would provide a feasible estimate of p for every value of x. Rather than modeling the variation in p with its limited range of $(0,1)$, we need to model the variation in some function of p that takes the full $(-\infty, \infty)$ range. Since the situation is binomial, the answer is provided by the natural parameter (see Chapter 5) of the binomial distribution, which is the logit. With a single continuous explanatory variable X the equivalently simple model would be

$$\ln\left(\frac{p}{1-p}\right) = \mu + \alpha x, \tag{7.1}$$

with x being the observed value of X. Note that when $p = 0$, $\ln\left(\frac{p}{1-p}\right) = -\infty$, and when $p = 1$, $\ln\left(\frac{p}{1-p}\right) = \infty$. Any intermediate value for $\mu + \alpha x$ can be converted into a value for p by exponentiating both sides of the equation and rearranging the terms to give

$$p = e^{\mu+\alpha x}/(1 + e^{\mu+\alpha x}). \tag{7.2}$$

Note that $p = 0.5$ corresponds to a zero value for the logit, with positive logit values corresponding to $p > 0.5$ and negative values to $p < 0.5$. A model of this type is described as a *logistic regression model*. The ratio $p/(1 - p)$ is the *odds* (on one outcome as opposed to the other) and thus an alternative name for the logit is the *log-odds*.

The relationship between p and the logit is illustrated in Figure 7.1. Note that, for p in the range $(0.2, 0.8)$, the values of its logit are nearly collinear. As a consequence, if the relation between p and some variable x appears to be approximately linear in this range, then the same will be true for the logit. The curved nature of the logit is only apparent when a wider range of values of p is considered.

7.2.1 Estimation of the Logit

Suppose that, in n trials we have r successes. The logit is the logarithm of the ratio of the proportion of successes to that of failures. A natural estimate of that ratio is therefore the logarithm of the number of successes divided by the number of failures: $\ln(r/(n - r))$. However, if p is very small (but greater than 0), then there will be a good chance that $r = 0$ which would mean

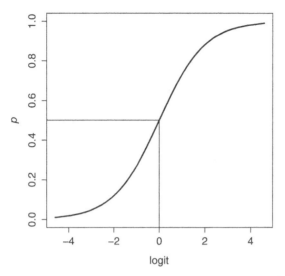

FIGURE 7.1 The relation between probability, p, and the logit, $\ln[p/(1-p)]$.

$\ln(r/(n-r)) = -\infty$. Similarly, if p is very large (but less than 1), then there will be a good chance that $r = n$ which would mean $\ln(r/(n-r)) = \infty$. In either case the true value of $\ln[p/(1-p)]$ would be finite and infinitely distant from the estimate! The solution proposed by Haldane (1956) was to add a small constant c to both observed frequencies. Since the choice $c = 0.5$ minimizes the bias of the estimator, Haldane's suggestion was to use:

$$\ln\left(\frac{r+0.5}{n-r+0.5}\right). \tag{7.3}$$

A comparison of alternative estimators is provided by Gart and Zweifel (1967).

7.2.2 The Null Model

We are interested in how the proportion of successes is affected by one or more explanatory variables. Suppose, however, that the proportion is not affected. Then the *null model* given by Equation (7.4) will be appropriate:

$$\ln\left(\frac{p}{1-p}\right) = \mu. \tag{7.4}$$

The goodness of fit of the null model provides a yardstick for comparison in assessing the performance of a more complex model. The degrees of freedom for this model will be one less than the number of observations (one, because the model has one parameter (μ) to estimate).

7.3 INDIVIDUAL DATA AND GROUPED DATA

Each individual data item takes one of two values: "Success" ($p = 1$) and "Failure" ($p = 0$). No intermediate values are possible. However, whatever the model, the predicted p-value for an individual data item will be intermediate between the extremes. No model of this type can provide a perfect explanation. As the example will show, a scatter diagram is unlikely to be helpful.

Example 7.1   Calcium intake and cardiovascular disease

The NHANES III cross-sectional study was conducted in the United States by the National Center for Health Statistics between 1988 and 1994. Over the following twelve years participants were monitored for survival or death. The data reported here (which refer to 16,052 individuals aged 17 or over who had no history of heart disease) were kindly made available by the authors of a study examining the possible link between average daily calcium intake and death due to cardiovascular disease (CVD).

The following R code creates Figure 7.3, which illustrates the data for a sample of 100 survivors and a sample of 100 who died of CVD.

```
R code
# The data are held in a vector named outcome
w1<-which(outcome==1);
w0<-which(outcome==0);
set<-c(1:100);
plot(data[c(w0[set],w1[set]),75],
    data[c(w0[set],w1[set]),40], lab=c(7,1,7),
    xlab="Mean daily total calcium consumption (in mg),
    x", ylab="Outcome (1 died of CVD; 0 other)")
```

With some imagination, the diagram hints that higher total calcium intake is associated with a reduced probability of dying from CVD. However (because of the stratified sampling) it does not show that only about 10% of the individuals in the trial died of CVD.

What Figure 7.2 has demonstrated is that plotting individual data is not going to be helpful. If there is sufficient data, then we can more easily see

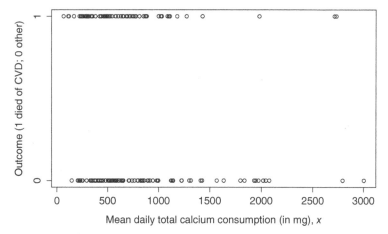

FIGURE 7.2   Scatter diagram showing the mean daily total calcium intake (in mg) for 100 individuals who died of cardiovascular disease (CVD), and for 100 who did not.

what is happening by grouping the data, and examining how the proportion (or the logit) varies for different ranges of *X*.

In Table 7.1, the *X*-variable (the estimated mean total daily calcium intake) has been subdivided into 10 categories. The category definitions were chosen so that there were comparable numbers in each category, with the category boundaries being rounded versions of the observed calcium percentiles (for example, 10% of individuals had an estimated calcium intake of less than 259 mg, so 250 mg has been used as the boundary of the first category).

Figure 7.3 shows that the proportion of individuals dying of CVD and the estimated total calcium intake do appear to be related. The two graphs (one

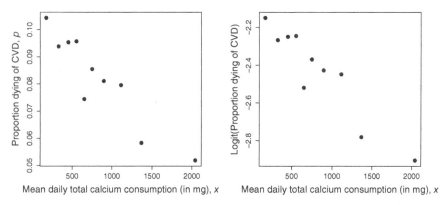

FIGURE 7.3   Graphs of mean daily total calcium intake (in mg) against proportion dying of CVD, and against logit(proportion dying of CVD).

TABLE 7.1 **Data from a long-term cohort study of the possible effect of calcium on cardiovascular disease (CVD)**

	Estimated mean total daily calcium intake, mg				
	< 250	250–	400–	500–	600–
Mean intake, x	173	327	451	552	649
Sample size, n	1476	2163	1646	1494	1397
Dying of CVD, k	154	203	157	143	104
% dying of CVD	10.4	9.4	9.5	9.6	7.4

	Estimated mean total daily calcium intake, mg				
	700–	800–	1000–	1250–	≥ 1500
Mean intake, x	748	895	1114	1366	2041
Sample size, n	1263	2034	1747	1097	1735
Dying of CVD, k	108	165	139	64	90
% dying of CVD	8.6	8.1	8.0	5.8	5.2

Source: The data were collected as part of the third National Health and Nutrition Examination Survey (NHANES III) and were kindly made available by the authors: van Hemelrijck, M., Michaelsson K., Linseisen J., Rohrmann S. (2013), *PLoS ONE* **8**(4): e61037. doi:10.1371/journal.pone.0061037.

for p and one for the logit) look very similar because of the short range of the values of p.

To apply the simple logistic regression model given by Equation (7.1) to the grouped data of Table 7.1, the following R code was used:

R code
```
calc<-c(173,327,451,552,649,748,895,1114,1366,2041);
n<-c(1476,2163,1646,1494,1397,1263,2034,1747,1097,1735);
die<-c(154,203,157,143,104,108,165,139,64,90);
mat<-cbind(die,n);
summary(glm(mat~calc,family=binomial))
```

The results are summarized in Table 7.2, which reports that

$$\ln\left(\frac{p}{1-p}\right) = -2.209 - 0.0003669x,$$

with the estimates of both μ and α being unquestionably different from zero. The second part of the output gives the values of G^2 (referred to as the

TABLE 7.2   Output (using R) for the fit of the simple logistic regression model, Equation (7.1), to the grouped data of Table 7.1

```
                Estimate Std. Error z value Pr(>|z|)
(Intercept) -2.209e+00    5.194e-02 -42.538 < 2e-16 ***
Calcium     -3.669e-04    5.916e-05  -6.202 5.57e-10 ***

   Null deviance: 46.5448 on 9 degrees of freedom
Residual deviance: 5.3833 on 8 degrees of freedom
```

deviance) for the simple logistic regression model ($G^2 = 5.3833$ with 8 d.f.) and for the null model (the model that omits αx). The difference between the two G^2-values provides another indication of the importance of including α in the model, since 41.1615 ($= 46.5448 - 5.3833$) would be an extraordinarily extreme value from a chi-squared distribution with $9 - 8 = 1$ d.f.

However, grouping the data was a device used for graphical purposes. It is the raw 0/1 data (partly illustrated in Figure 7.3) to which the model should be applied. The following R commands

---

R code (*continued*)

```
model<-glm(outcome~calc,family=binomial);
summary(model)
```

---

result in the output summarized in Table 7.3. The estimates there are slightly different to those for the grouped data: we now have $\hat{\mu} = -2.10$, and $\hat{\alpha} = -0.00039$. For a person with an average daily total calcium intake of x mg, the probability of dying from CVD in a 12-year period (the original data referred to such a period) would be estimated to be

$$e^{-2.1-0.00039x} / \left(1 + e^{-2.1-0.00039x}\right).$$

The fitted relation is illustrated in Figure 7.4.

TABLE 7.3   Output (using R) for the fit of the simple logistic regression model to the 16,052 observations summarized in Table 7.1

```
                Estimate Std. Error z value Pr(>|z|)
(Intercept) -2.103e+00    5.178e-02 -40.606 < 2e-16 ***
Calcium     -3.935e-04    5.917e-05  -6.651 2.91e-11 ***

   Null deviance: 9157.3 on 16051 degrees of freedom
Residual deviance: 9108.1 on 16050 degrees of freedom
```

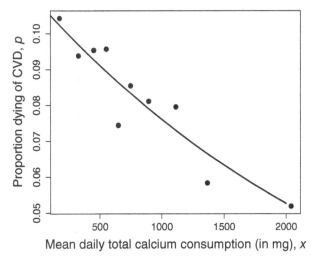

FIGURE 7.4 The probabilities derived from the fitted simple logistic regression model superimposed on the scatter diagram of mean daily total calcium intake (in mg) against proportion dying of CVD.

All the results have shown that the proportion dying of CVD is lower for those with a high mean daily calcium intake. A naive interpretation would be that one can avoid CVD by increasing the calcium intake. However, there are several reasons why this may be untrue: foremost amongst these is the possibility that high levels of calcium may lead to death by other causes (before CVD can take effect).

7.4 PRECISION, CONFIDENCE INTERVALS, AND PREDICTION INTERVALS

Parameter estimation is useful, but quite how useful will depend on the precision of that estimate. Fortunately, whichever computer program is used, it is sure to provide appropriate information. With large samples, the parameter estimates are approximately normally distributed, so that for a parameter α, with estimated value $\hat{\alpha}$ and estimated standard error $se(\hat{\alpha})$ (the square root of the estimated variance of $\hat{\alpha}$) the ratio z, given by

$$\hat{\alpha}/se(\hat{\alpha})$$

has an approximate standard normal distribution. It follows that an approximate 95% confidence interval is provided by

$$\hat{\alpha} \pm 1.96se(\hat{\alpha}). \tag{7.5}$$

This interval may be referred to as a *Wald confidence interval* after the Hungarian statistician Abraham Wald (1902–1950).

The Wald interval is useful as a quick approximation, but a more accurate interval is provided by studying how the likelihood changes as the value suggested for the parameter is varied. For a single parameter, a reduction in the likelihood greater than $1.96^2 = 3.84$ would be significant at the 5% level, since the reference distribution would be a chi-squared distribution with one degree of freedom. Most computer programs will do the hard work and report these likelihood-based intervals.

Example 7.1   Calcium and cardiovascular disease (*continued*)

Continuing with the analysis given in Table 7.3, we can obtain the Wald and likelihood-based 95% confidence intervals for $\hat{\alpha}$ using these R commands:

R code (*continued*)

```
confint.default(model);              # Wald intervals
confint(model);            # Likelihood based intervals
```

Table 7.4 gives results both for the full set of 16,052 observations, and for a random sample of 180 observations. The table shows that, for very large samples, the Wald interval is a good representation of the more precise likelihood-based interval. However, for smaller numbers of observations, the Wald interval may mislead the analyst: in this example the Wald interval for the sample of 180 observations spans zero and therefore incorrectly implies that α does not differ significantly from zero at the 5% level.

7.4.1 Prediction Intervals

The estimated value of a logit is calculated as a linear combination of the estimated parameters in the model. Each of these estimates has its own variance and, in general, the estimates will be correlated (i.e., there will be non-zero covariances). It follows that it would be difficult to estimate the variance of

TABLE 7.4   Wald, and likelihood-based, 95% confidence intervals for the apparent effect of calcium intake on the probability of death

	All 16,052 observations	Random sample of 180 observations
Wald interval	(−0.00051, −0.00028)	(−0.00377, 0.00009)
Likelihood-based interval	(−0.00051, −0.00028)	(−0.00407, −0.00022)

an estimated logit by hand. Fortunately, we don't need to, since the computer finds the hard work easy. Assume that, for a given value of the explanatory variable, the computer reports the estimated logit, $\hat{l}$ and its standard error, $se(\hat{l})$. An approximate 95% prediction interval for the logit is therefore

$$\hat{l} \pm 1.96se(\hat{l}),$$

so that an approximate 95% prediction interval for the success probability p is given by

$$\left(\frac{\exp\left\{\hat{l} - 1.96se(\hat{l})\right\}}{1 + \exp\left\{\hat{l} - 1.96se(\hat{l})\right\}}, \quad \frac{\exp\left\{\hat{l} + 1.96se(\hat{l})\right\}}{1 + \exp\left\{\hat{l} + 1.96se(\hat{l})\right\}} \right). \tag{7.6}$$

Example 7.1   Calcium and cardiovascular disease (*continued*)

Figure 7.5 shows the result of using the bounds given by Equation (7.6) to the simple logistic model given by Equation (7.1) applied to the CVD data. Notice the way that the bounds widen away from the center of the data.

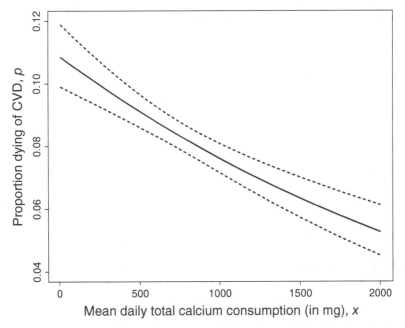

FIGURE 7.5 The approximate 95% prediction interval for the simple logistic regression model, Equation (7.1), relating CVD to calcium intake.

7.5 LOGISTIC REGRESSION WITH A CATEGORICAL EXPLANATORY VARIABLE

Often an explanatory variable is categorical rather than continuous. Common examples are gender and ethnicity. Since the response variable might behave in a different fashion for each gender, or each ethnic group, the model will need separate parameters for every category. For an explanatory variable with J categories, a simple solution would be

$$\ln\left(\frac{p}{1-p}\right) = \mu_j, \qquad j = 1, 2, \ldots, J. \tag{7.7}$$

However, this is by no means the only way of writing the model. One alternative is

$$\ln\left(\frac{p}{1-p}\right) = \mu + \alpha_j, \qquad j = 2, \ldots, J. \tag{7.8}$$

Here, the first of the categories is a reference category with which each of the other categories is compared. Thus, if α_j does not differ significantly from zero, then this implies that categories 1 and j do not differ significantly. This is one choice of model often selected by computer programs, while another is to use the last of the J categories as the reference category:

$$\ln\left(\frac{p}{1-p}\right) = \mu + \alpha_j, \qquad j = 1, \ldots, J - 1. \tag{7.9}$$

A third alternative compares the effect of each category with that of an "average category":

$$\ln\left(\frac{p}{1-p}\right) = \mu + \alpha_j \qquad j = 1, \ldots, J, \qquad \sum_{j=1}^{J} \alpha_j = 0. \tag{7.10}$$

Whichever of the three forms of model is used, if the explanatory variable is irrelevant, then all α-parameters will be zero.

Example 7.2 Gender and cardiovascular disease

For the NHANES III data analyzed previously, information was collected on many potential explanatory variables. Table 7.5 reports the incidence of CVD deaths separately for males and females. The logits in the table have been calculated using Equation (7.3).

TABLE 7.5   The incidence of CVD deaths for males and females during the 12-year follow-up period of the NHANES III study

	Males	Females
Deaths from CVD	670	657
Other	6735	7990
Logit	−2.31	−2.50

In order to save storage space most studies use codes in their data summary. An example is the use of "1" for "Male" and "2" for "Female". In this case, therefore, 2 is not twice 1; it is simply an alternative to 1. The computer program needs to know that that is so. The following R code gives the necessary commands.

R code

```
# A vector named Gender exists
Gender<-as.factor(Gender);
summary(glm(outcome~Gender,family=binomial))
```

The resulting output is shown in Table 7.6. The first part of the output refers to Gender2 which is the second gender category (Female). Since the first category is not mentioned, it is the reference category and the output is using the form of model given by Equation (7.8). For $j = 1$, the equation becomes

$$\ln\left(\frac{p}{1-p}\right) = \mu,$$

TABLE 7.6   Output (using R) for the fit of the simple logistic regression model with gender as the single explanatory variable

```
            Estimate Std. Error z value Pr(>|z|)
(Intercept) -2.30780    0.04051 -56.969  < 2e-16 ***
Gender2     -0.19047    0.05734  -3.322 0.000895 ***

   Null deviance: 9157.3  on 16051  degrees of freedom
Residual deviance: 9146.3  on 16050  degrees of freedom
```

so that the intercept μ corresponds to the logit for the first category. In this case, therefore, the estimated logit for males is -2.30780 and that for females is

$$-2.30780 - 0.19047 = -2.49827.$$

These are the values reported to two decimal places (which is quite enough!) in Table 7.5. That the difference is highly significant is indicated by the tail probability (0.000895).

The second part of the table contrasts the goodness-of-fit of the null model given by Equation (7.4), with that of the model given by Equation (7.8). There is a reported difference of $9157.3 - 9146.3 = 11.0$ for a change of one degree of freedom (more accurately, the difference is 11.033). The probability that a χ_1^2-distribution takes the value 11.033, or a more extreme value, is 0.000895, as reported in the first part of the table.

Note that, since $J = 2$, the result is a 2×2 table for which the methods of Chapter 3 are also appropriate.

7.5.1 Parameter Estimates with Categorical Variables (*J* > 2)

For a categorical variable with J categories, computer programs will economically report just $(J - 1)$ parameter estimates. The estimates reported will depend upon the formulation. An example is provided by Table 7.7 which refers to a variable having categories (Red, White, Blue).

TABLE 7.7   The reported parameter estimates for a three-category variable using alternative formulations of the same model

Equation (7.8)		Equation (7.9)		Equation (7.10)	
		Red	−3	Red	−1
White	0	White	−3	White	−1
Blue	3				

In the first formulation, the first category Red is the reference category. The zero value for White implies equal estimates for Red and White, with Blue being three greater.

For the second formulation, the last category Blue is the reference category. The estimates for Red and White are reported as being three less than that for Blue.

The third formulation uses the average category as reference. To interpret the estimates for the final formulation (7.10) we need to determine the missing value; in this formulation the estimates sum to zero, so the missing value is two. Thus the value for Blue is again three greater than each of the other two estimates. The need for care in interpretation is very evident.

Example 7.3   Cardiovascular disease and ethnicity

In the NHANES III study ethnicity was recorded using four categories (1: Non-Hispanic white; 2: Non-Hispanic black; 3: Mexican American; 4: Other). The proportions dying of cardiovascular disease vary noticeably according to ethnicity, as Table 7.8 shows.

TABLE 7.8   The numbers dying of cardiovascular disease, subdivided by ethnicity

	1: Non-Hispanic white	2: Non-Hispanic black	3: Mexican American	4: Other
Dying of CVD	764	318	219	26
Other	5542	4306	4259	618
Percentage dying of CVD	12.1	6.9	4.9	4.0

Once again, when we fit the model (in the form of Equation 7.8) the estimates (see Table 7.9) correspond to the observed logits. Thus $\ln(764/5542) = -1.9814$, $\ln(318/4306) = (-1.98154 - 0.06217) = -2.60571$, and so on. The differences between group 1 and each of ethnic groups 2 to 4 are highly significant.

TABLE 7.9   Output (using R) for the fit of the simple logistic regression model with ethnicity as the single explanatory variable

```
             Estimate Std. Error z value Pr(>|z|)
(Intercept) -1.98154    0.03859 -51.346  < 2e-16 ***
Ethnic2     -0.62417    0.06976  -8.948  < 2e-16 ***
Ethnic3     -0.98617    0.07931 -12.435  < 2e-16 ***
Ethnic4     -1.18685    0.20384  -5.823 5.79e-09 ***

     Null deviance: 9157.3  on 16051  degrees of freedom
 Residual deviance: 8939.5  on 16048  degrees of freedom
```

7.5.2 The Dummy Variable Representation of a Categorical Variable

The model represented by any of Equations (7.7)–(7.10) has J unknown parameters, though this may not be immediately apparent. An alternative is to use so-called *dummy variables*. A dummy variable takes the value 1 if some condition is true, and otherwise takes the value 0. An alternative description is *indicator variable*.

We will define the dummy variable D_j as follows:

$$D_j = \begin{cases} 1 & \text{if an individual belongs to category } j, \\ 0 & \text{otherwise.} \end{cases} \tag{7.11}$$

Using these dummy variables Equation (7.7) can be written as

$$\ln\left(\frac{p}{1-p}\right) = \alpha_1 D_1 + \alpha_2 D_2 + \cdots + \alpha_J D_J, \tag{7.12}$$

while Equation (7.8) can be written as

$$\ln\left(\frac{p}{1-p}\right) = \mu + \alpha_2 D_2 + \cdots + \alpha_J D_J, \tag{7.13}$$

with equivalent forms for Equations (7.9) and (7.10).

REFERENCES

Gart, J. J., and Zweifel, J. R. (1967) On the bias of various estimators of the logit and its variance. *Biometrika*, **54**, 181–187.

Haldane, J. B. S. (1956) The estimation and significance of the logarithm of a ratio of frequencies. *Ann. Human Genet.*, **35**, 297–303.

CHAPTER 8

LOGISTIC REGRESSION WITH SEVERAL EXPLANATORY VARIABLES

Logistic regression is not confined to a single explanatory variable, nor is it necessary for the explanatory variables to be all of the same type.

8.1 DEGREES OF FREEDOM WHEN THERE ARE NO INTERACTIONS

All computer programs will report the number of degrees of freedom for a given model. That number is given by

d.f. = Number of observations − Number of parameters.

The number of parameters is easily determined. For example, for a model with two continuous variables and two categorical variables:

Overall mean	1
Each continuous variable	1
Categorical variable (I categories)	$I - 1$
Categorical variable (J categories)	$J - 1$

Categorical Data Analysis by Example, First Edition. Graham J. G. Upton.
© 2017 John Wiley & Sons, Inc. Published 2017 by John Wiley & Sons, Inc.

With two categorical explanatory variables (with I and J categories) and with two continuous explanatory variables (X and Y), a simple model would be:

$$\ln\left(\frac{p}{1-p}\right) = \mu + \alpha_i + \beta_j + \gamma x + \delta y, \quad i = 2, \ldots, I, j = 2, \ldots, J. \quad (8.1)$$

In this case, with n observations there would be

$$n - 1 - (I - 1) - (J - 1) - 1 - 1 = n - I - J - 1$$

degrees of freedom.

Example 8.1   A four-variable model for the CVD data

We now bring together all the explanatory variables previously considered into a single model:

$$\ln\left(\frac{p_{ij}}{1-p_{ij}}\right) = \mu + \text{Gender}_i + \text{Ethnicity}_j + \gamma\text{Age} + \delta\text{Calcium},$$

$$i = 1, \ldots, I, j = 1, \ldots, J.$$

An appropriate R command is

R code (*continued*)

```
# The vectors Age, Calcium, Gender, Ethnic exist
summary(glm(outcome~Age+Calcium+Gender+Ethnic,
    family=binomial))
```

By default R takes $\text{Gender}_1 = 0$ and $\text{Ethnicity}_1 = 0$; the resulting output is given in Table 8.1.

In this case there are $1 + (2 - 1) + (4 - 1) + 1 + 1 = 7$ parameters; with 16,052 observations that leaves 16,045 degrees of freedom.

The interpretation of the estimated values of the ethnicity parameters is that there is no significant difference between Mexican American (the third ethnicity category) and Non-Hispanic white (the reference category for ethnicity). The major ethnicity difference is the much greater value (corresponding to an increased probability of CVD death) for Non-Hispanic blacks (the second ethnicity category) as opposed to Non-Hispanic whites. Age is again the dominant effect, but there continues to be a small, but significant, effect due to calcium intake.

TABLE 8.1 Output (using R) for the fit of the logistic regression model that includes the four-category ethnicity variable

	Estimate	Std. Error	z value	Pr(>|z|)	
(Intercept)	-7.604e+00	1.944e-01	-39.122	< 2e-16	***
Age	9.308e-02	2.416e-03	38.523	< 2e-16	***
Calcium	-1.977e-04	6.852e-05	-2.885	0.00392	**
Gender2	-2.673e-01	6.569e-02	-4.069	4.72e-05	***
Ethnic2	3.363e-01	8.397e-02	4.005	6.21e-05	***
Ethnic3	4.784e-02	9.131e-02	0.524	0.60034	
Ethnic4	-4.606e-01	2.234e-01	-2.062	0.03923	*

```
    Null deviance: 9157.3  on 16051  degrees of freedom
Residual deviance: 6355.5  on 16045  degrees of freedom
```

8.2 GETTING A FEEL FOR THE DATA

When there are several variables, it is usually worth constructing figures and tables that demonstrate how the probability of a "success" varies across different combinations of the explanatory variables. Simple diagrams are often revealing. Table 8.2 gives an indication of the sorts of issues that may become apparent as a result. The whole process might be described as "getting a feel for the data."

TABLE 8.2 Problems that may be revealed by a detailed inspection of the data using tables and plots

Problem	Solution
Non-linearity	Add polynomial terms, or transform the explanatory variable (e.g., take logarithms or reciprocals)
Trend	Trend across two variables implies their interaction may be required
Outliers	Determine cause; perhaps add extra dummy variable
Data inconsistencies	Check for mistyping and programming errors

Example 8.2   CVD, calcium intake, and gender

It might be that the apparent effect of gender on the CVD death rates (or, alternatively, the apparent effect of calcium on the CVD deaths) was due to some difference in the calcium intake of the two genders. To investigate this idea, we can work with the grouped data and examine the CVD death rates separately for males and females. Figure 8.1 shows the result; it indicates that there is a gender effect that is approximately constant across the range of calcium consumption.

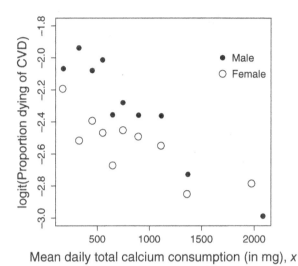

FIGURE 8.1 Scatter diagram showing the dependence of the logit(dying of CVD) on mean daily total calcium intake and gender.

Example 8.3 CVD, calcium intake, and age

Any consideration of the probability of CVD death should take account of an individual's age. Disregarding age might well lead us to mistaken conclusions concerning the relevance of other variables. For example, if calcium intake was much higher amongst the young than the old (perhaps because older people eat less) then high calcium intake could appear to be associated with low CVD incidence for no reason other than variations in appetite! With the large amount of data that are available in this study, we can examine the variation due to calcium intake separately for different age brackets. Table 8.3

TABLE 8.3 Variations in percentages dying from CVD: data subdivided by age and calcium intake

Age range	Estimated mean total daily calcium intake, mg								
	< 250	250–	400–	500–	600–	700–	800–	1000–	≥ 1250
< 50	1	1	1	1	1	2	2	1	1
50–	9	7	6	7	6	3	2	5	3
60–	15	17	9	9	14	13	10	10	8
65–	18	21	24	23	23	22	18	21	14
75–	47	45	42	50	35	41	47	43	42
Overall	10	9	10	10	7	9	8	8	5

shows the result. Controlling for age (by looking across the rows of the table), there does appear to be a slight reduction in the proportions dying from CVD as the calcium intake increases, though, since the decrease is not a steady one, this may suggest the need to add a calcium-age interaction to the model.

8.3 MODELS WITH TWO-VARIABLE INTERACTIONS

All the models so far considered have been purely additive. Often, however, the combined effect of two variables is different to the sum of their separate effects. For example, one cannot hear an empty balloon being pricked by a pin, nor an empty balloon being filled with air, but a filled balloon pricked by a pin may well go bang! We can capture this idea mathematically by including terms that involve the product of the explanatory variables. With two continuous explanatory variables (X and Y), the model with interaction is

$$\ln\left(\frac{p}{1-p}\right) = \mu + \alpha x + \beta y + \theta xy. \tag{8.2}$$

This model simplifies to the additive model if $\theta = 0$.

If one explanatory variable (X) is categorical (with I categories) then the model becomes

$$\ln\left(\frac{p}{1-p}\right) = \mu + \alpha_i + \beta y + \theta_i y, \qquad i = 2, \dots, I. \tag{8.3}$$

Thus, for the first category of X the right-hand side of the equation is $\mu + \beta y$, while, for the second category, the intercept (quantifying the underlying effect of that category of X) changes to $\mu + \alpha_2$ and the slope (quantifying the dependence on the continuous variable Y) changes to $\beta + \theta_2$.

Example 8.4   Including the gender-age interaction in modeling the incidence of CVD

Previous analyses have demonstrated that the probability of death from CVD varies with age and gender. Incorporating an age-gender interaction allows for an investigation of whether age affects the two genders in a subtly different fashion so far as death from CVD is concerned.

An appropriate R command to investigate the fit of the interaction model given by Equation 8.3 is:

R code (*continued*)

```
summary(glm(outcome~Age*Gender,family=binomial))
```

The resulting output is given in Table 8.4.

TABLE 8.4 Output (using R) for the fit of the logistic regression model that includes the gender-age interaction

```
              Estimate Std. Error  z value  Pr(>|z|)
(Intercept) -5.576133   0.476056  -11.713   < 2e-16 ***
Age          0.065537   0.006935    9.450   < 2e-16 ***
Gender      -1.445531   0.320068   -4.516  6.29e-06 ***
Age:Gender   0.017587   0.004594    3.828  0.000129 ***


    Null deviance: 9157.3  on 16051  degrees of freedom
Residual deviance: 6378.4  on 16048  degrees of freedom
```

The table shows that there is indeed a significant interaction with

$$\ln\left(\frac{p}{1-p}\right) = -5.58 + 0.0655x$$

for males, and

$$\ln\left(\frac{p}{1-p}\right) = (-5.58 - 1.45) + (0.0655 + 0.0176)x = -7.02 + 0.0831x$$

for females.

The second diagram in Figure 8.2 illustrates these relations, with the first diagram showing the fitted curves that arise when the interaction term is omitted. The purely additive model shows a consistently higher probability for males across the entire age range. Introducing the interaction term allows for response curves with different shapes: it is now apparent that amongst younger respondents the probability of death from CVD is higher for males, whereas for older respondents it is females that are most at risk from the disease.

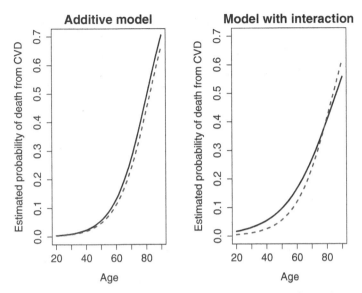

FIGURE 8.2   The estimated probabilities of dying from CVD for males (solid line) and females (dashed line) for the purely additive model, using age and gender, and for the model including the age-gender interaction.

8.3.1 Link to the Testing of Independence Between Two Variables

If both explanatory variables are categorical (with X having I categories and Y having J categories) then the model including an interaction is

$$\ln\left(p_{ij}/(1 - p_{ij})\right) = \mu + \alpha_i + \beta_j + \theta_{ij}, \quad i = 2, \dots, I, \quad j = 2, \dots, J, \quad (8.4)$$

where p_{ij} is the probability of a success for an observation belonging to category i of X and category j of Y. In this equation there are $(I - 1)$ parameters associated with the main effect of X and $(J - 1)$ parameters associated with the main effect of Y, so that there are $(I - 1)(J - 1)$ parameters associated with the XY interaction. With no interaction between X and Y, the model would simplify to

$$\ln(p_{ij}/(1 - p_{ij})) = \mu + \alpha_i + \beta_j, \quad i = 2, \dots, I, \quad j = 2, \dots, J. \quad (8.5)$$

This simplified model, which states that the effects of X and Y are independent, has $(I - 1)(J - 1)$ fewer parameters than its predecessor. A test of whether the interaction is required (i.e., a test of the hypothesis that the two variables are independent) is provided by comparing the fits of the two models. The difference in the goodness-of-fit values will be linked to the $(I-1)(J-1)$ omitted parameters and will therefore have $(I-1)(J-1)$ degrees of freedom. This is why the chi-squared tests of independence in Chapter 4 were associated with that number of degrees of freedom.

CHAPTER 9

MODEL SELECTION AND DIAGNOSTICS

This chapter is concerned with modeling data and examining the fit in the context of logistic regression. However, the general principles are not confined to this type of model; they apply equally well to multiple regression models with continuous random variables, and to the log-linear models discussed in subsequent chapters.

9.1 INTRODUCTION

In science we encounter many laws that are equations linking measurable properties. For example, Boyle's law states that the pressure exerted by a gas is inversely related to the volume it occupies, while Avogadro's law states that, under constant conditions, the volume of a gas is proportional to the amount of substance present. Any apparent deviations from these laws can be ascribed to measurement inaccuracy. In these cases there is no doubt concerning which variables are relevant, nor about the form of the model. However, in the social sciences (and elsewhere in science, to be fair) these certainties do not exist; we are often not sure which variables are relevant, nor how they are related. In this chapter, these issues are addressed in the context of logistic regression, though the approaches used have more general relevance.

Categorical Data Analysis by Example, First Edition. Graham J. G. Upton.
© 2017 John Wiley & Sons, Inc. Published 2017 by John Wiley & Sons, Inc.

9.1.1 Ockham's Razor

William of Ockham was a Franciscan friar believed to have been born in Ockham (Surrey, England) in about 1287. He is now best remembered for his proposition that, when there are alternative explanations of some phenomenon, it is the simpler explanation that should be preferred; this is the principle now called Ockham's razor or the *principle of parsimony*.

In the modeling context this implies that, if two models provide equally good explanations of the data, then the model with the fewer unknown parameters is the preferred model (providing it makes good sense to the investigator).

Example 9.1 Birth months of mathematicians

The MacTutor website provides biographies of famous mathematicians. It also provides, for every day of the year, the numbers of persons in their archive who were born on that day. The data are summarized in Table 9.1, together with the numbers for biographies containing the words "statistician" or "probabilist." Evidently there are more famous mathematicians born in some months than in others (your author was born in January, but declines to comment further). Our interest is in the possibility that the words "statistician" or "probabilist" are disproportionately more likely to appear in the biographies of mathematicians born in certain months (the data suggest February and September) than others (e.g., July). The null hypothesis is that the proportion is constant across the months. Using the Pearson X^2 goodness-of-fit statistic (Equation 2.1) gives $X^2 = 17.88$ with 11 degrees of freedom. Since the tail probability is greater than 10%, this is not an unusually large value and the null hypothesis is therefore accepted.

TABLE 9.1   Monthly counts of birthdays of mathematicians with biographies on the MacTutor website, together with numbers of biographies that include one or more of the words "statistician" or "probabilist"

Month	Jan	Feb	Mar	Apr	May	Jun
All biographies	201	195	192	194	187	191
Statistician/probabilist	5	9	3	5	1	4

Month	Jul	Aug	Sep	Oct	Nov	Dec
All biographies	160	166	190	167	150	170
Statistician/probabilist	0	7	8	3	6	3

Source: http://www-history.mcs.st-and.ac.uk/Miscellaneous/b_d_stats.html

The constant-proportion model has the benefit of simplicity, but it is not a perfect explanation of the data. A perfect description is available, but it is not simple. This description, which uses 11 extra parameters (using up the 11 degrees of freedom), states that the proportion for January is 5/201 and the proportion for February is 9/195, etc. In the present context, since the constant-proportion model is both a simpler model and is easier to understand (and more believable), it seems preferable.

As a codicil we note that, in this case, since the expected numbers in the statistician/probabilist category are very small, the χ^2 approximation is a little suspect. An application of the exact test (introduced earlier in the context of a 2×2 table) suggests a tail probability of 5.1%.

9.2 NOTATION FOR INTERACTIONS AND FOR MODELS

With two continuous explanatory variables the model that included their interaction was given by Equation (8.2) as

$$\ln\left(\frac{p}{1-p}\right) = \mu + \alpha x + \beta y + \theta xy.$$

With three explanatory variables (X, Y, and Z) we might need to include two more two-variable interactions and also the three-variable interaction (which would imply that each two-variable interaction varied with the value of third variable). This model would have 2^3 unknown parameters. Similarly, with four continuous explanatory variables, there could be 2^4 unknown parameters.

With categorical variables having more than two categories the number of unknowns would be greater still. For example, with three categorical variables having 2, 3, and 4 categories, there could be as many as 24 unknowns in the model. Rather than using long equations, we will use the compact notation illustrated in Table 9.2.

In each case the shorthand description consists of a list of the most complex interactions, with the understanding that if a complex interaction is included in the model, then so must all the simpler interactions that form part of its "construction." Thus XYZ implies that the XY, XZ, and YZ interactions are included, while XY similarly implies that the main effects of X and Y must be included. These hierarchical constraints are a consequence of Birch's result (Section 13.1).

A corollary is that some models will not be considered. One such is

$$\ln\left(\frac{p}{1-p}\right) = \mu + \alpha x + \gamma z + \theta xy.$$

TABLE 9.2 Shorthand notation used for models of varying complexity

Model	Shorthand
Models with two explanatory variables	
$\ln\left(\frac{p}{1-p}\right) = \mu + \alpha x$	X
$\ln\left(\frac{p}{1-p}\right) = \mu + \alpha x + \beta y$	X/Y
$\ln\left(\frac{p}{1-p}\right) = \mu + \alpha x + \beta y + \theta xy$	XY
Models with three explanatory variables	
$\ln\left(\frac{p}{1-p}\right) = \mu + \alpha x + \beta y + \gamma z$	$X/Y/Z$
$\ln\left(\frac{p}{1-p}\right) = \mu + \alpha x + \beta y + \gamma z + \theta xy$	XY/Z
$\ln\left(\frac{p}{1-p}\right) = \mu + \alpha x + \beta y + \gamma z + \theta xy + \phi xz$	XY/XZ
$\ln\left(\frac{p}{1-p}\right) = \mu + \alpha x + \beta y + \gamma z + \theta xy + \phi xz + \rho yz$	$XY/XZ/YZ$
$\ln\left(\frac{p}{1-p}\right) = \mu + \alpha x + \beta y + \gamma z + \theta xy + \phi xz + \rho yz + \tau xyz$	XYZ

Because this model includes the interaction term θxy, it should also include the main effect term βy. The underlying argument is that, if the combination of particular values of X and Y matters, then that implies that both the particular values of X matter and the particular values of Y matter. If the particular values of Y matter, then βy should be included in the model. A helpful side-effect is that it can greatly reduce the number of different models that need to be considered.

9.3 STEPWISE METHODS FOR MODEL SELECTION USING G²

Table 9.2 presented many of the possible models for the case of three explanatory variables, but there are others. The complete tree of models is illustrated in Figure 9.1. In the tree, any pair of models connected by a line differ by just one parameter or set of parameters (depending on whether continuous or categorical variables are involved).

The number of possible models rises rapidly as the number of variables grows, so that the selection of a best (in some sense) model is a major challenge when there are many explanatory variables. However, Figure 9.1 suggests a way for proceeding that is based on the comparison of nested models.

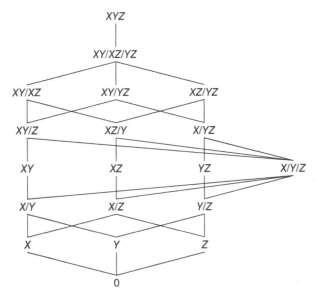

FIGURE 9.1 The tree of possible models involving three explanatory variables, X, Y, and Z.

Suppose that M_1 is a model with d_1 degrees of freedom and lack-of-fit measured by G_1^2. Similarly, let M_2 be a model with d_2 degrees of freedom and lack-of-fit given by G_2^2. If M_2 includes all the parameters that were in model M_1, together with those in the set S, then G_2^2 cannot be greater than G_1^2, since the addition of extra explanatory terms cannot result in a worsening of the fit. It follows that $G_1^2 - G_2^2$, the reduction in the lack of fit, must be a direct consequence of including the parameters in the set S. The significance of that reduction can be assessed by comparison with a chi-squared distribution with $d_1 - d_2$ degrees of freedom. Some examples of pairs of models differing by a single set of parameters are given in Table 9.3.

There are three related approaches to model selection based on model pairs: forward selection, backward elimination, and complete stepwise.

TABLE 9.3 **Pairs of models differing by a single group of parameters in a case where X has I categories, Y has J categories, and Z is a continuous variable**

Simpler model	More complex model	Difference	Change in d.f.
Y	Y/Z	Z	1
Y/Z	$X/Y/Z$	X	$I-1$
$X/Y/Z$	XY/Z	XY	$(I-1)(J-1)$

9.3.1 Forward Selection

Forward selection consists of adding successive sets of parameters to the current model. As an example, suppose that the current model is X/Z. Figure 9.1 indicates that there are two potential next models: XZ and $X/Y/Z$. With forward selection we select whichever is the more successful. Suppose that this is $X/Y/Z$. At the next step, as the figure shows, there are three candidate models: XY/Z, XZ/Y, and X/YZ. Again we choose whichever is the best. This process continues until the change in G^2 appears not to be significant.

Example 9.2 The UK 1975 referendum

In 1975, British voters were invited to vote for, or against, Britain's entry into the Common Market. There were three official leaflets distributed to voters before the referendum, so one variable considered here (denoted by R) is whether or not a voters read at least one of those leaflets. A second variable D concerns when a voter arrived at a decision: its three categories are "A long time ago," "Sometime this year," and "Only a little before the referendum." The third variable (undoubtedly important) was P, the political affiliation of the voter (Conservative, Labour, Liberal). A selection from the raw data is given in Table 9.4.

TABLE 9.4 A selection from data referring to the 1975 UK referendum

Respondent	Vote	Read leaflet (R)	Made decision (D)	Party supported (P)
1	For	At least one	Sometime this year	Conservative
2	Against	At least one	Sometime this year	Labour
⋮	⋮	⋮	⋮	⋮
1453	Against	At least one	Only a little before	Labour
1454	Against	At least one	Sometime this year	Labour

Source: SN 830, British Election Study: EEC Referendum Survey, 1975. Reproduced with permission of the UK Data Service.

Figure 9.2 shows the route that would be taken if forward selection was taken to its limit. A total of 13 models would be examined while the fit of 6 models would not be tested. In practice, forward selection would finish after fitting the nine models listed in Table 9.5 (from which the decision process can be inferred). For example, having reached the model P/D, the decision to add the PD interaction results from observing that the reduction in G^2 from 1492.0 to 1470.7, which adds four extra parameters, is a reduction that would occur by chance with a tail probability of about 0.0003.

The final model selected is the model PD/R, since the possible additions to this model do not make significant improvements.

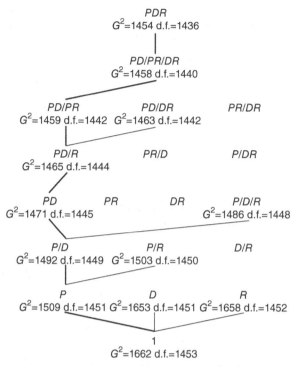

FIGURE 9.2   Forward selection for the referendum data. Models examined are indicated, with the entire preferred path shown in bold. Key: P, political affiliation; D, time of decision; R, read official leaflet.

TABLE 9.5   Forward selection through the referendum data (selected models in bold)

Model	G^2	d.f.	Change in model	Change in G^2	d.f.	Tail probability
1	1662.2	1453				
P	1509.0	1451	*P*	153.2	2	0
P/D	1492.0	1449	*D*	17.0	2	0.0002
P/R	1503.3	1450	*R*	5.7	1	0.017
PD	1470.7	1445	*PD* interaction	21.3	4	0.0003
P/D/R	1486.0	1448	*R*	6.0	1	0.014
PD/R	1464.9	1444	*R*	5.8	1	0.016
PD/PR	1459.2	1442	*PR* interaction	5.7	2	0.057
PD/DR	1463.5	1442	*DR* interaction	1.4	2	0.497

Key: P, political affiliation; D, time of decision; R, read official leaflet.

TABLE 9.6 Observed proportions (in italics) voting in favor of entry to the Common Market, together with the estimated proportions using the model PD/R and, in brackets, the model $P/D/R$. Estimates are for those who read at least one leaflet

Party	When decided on voting in favor								
	A long time ago			Sometime this year			Only a little before		
Conservative	*0.87*	0.87	(0.84)	*0.91*	0.93	(0.91)	*0.76*	0.79	(0.89)
Labour	*0.43*	0.45	(0.48)	*0.60*	0.60	(0.63)	*0.70*	0.66	(0.59)
Liberal	*0.69*	0.66	(0.69)	*0.88*	0.85	(0.80)	*0.80*	0.77	(0.77)

In Table 9.5 the smaller tail probabilities are associated with the PD interaction, which therefore needs examination. Table 9.6 shows, for those who read at least one leaflet, the observed proportions voting in favor for each combination of these two variables together with the estimated proportions according to the models PD/R and $P/D/R$. Comparing the observed and estimated proportions, it can be seen that for most category combinations the $P/D/R$ model gives good results. However, amongst Labour supporters who made a decision just before voting, the proportion voting in favor (70%) was much higher than predicted by the $P/D/R$ model.

Comparisons of estimates with and without a particular term being included in the model are always a good way of seeing the importance of the term.

9.3.2 Backward Elimination

This is the same idea as forward selection, but in reverse. From a relatively complex model, that provides an acceptable fit to the data (as measured by G^2), we attempt a simplification by eliminating a set of parameters. At each step, we choose to eliminate whichever set of parameters appears to be of least significance. Successive eliminations occur until it is not possible to find a simpler model that provides an adequate explanation of the data.

Example 9.2 The UK 1975 referendum (*continued*)

Figure 9.3 shows that, in this particular case, the optimal path is the same in each direction. The same final model would be chosen (though only after all 13 models had been tested). In other data sets, with a greater number of explanatory variables, the paths can be very different with different final models being selected. The values of G^2 and the tail probabilities associated with the changes in G^2 are shown in Table 9.7.

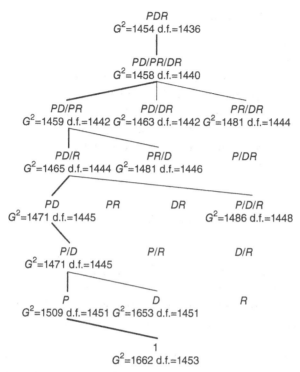

FIGURE 9.3 Backward elimination for the referendum data. Models examined are indicated, with the entire preferred path shown in bold. Key: *P*, political affiliation; *D*, time of decision; *R*, read official leaflet.

TABLE 9.7 **Backward elimination with the referendum data (selected models in bold)**

Model	G^2	d.f.	Change in model	Change in G^2	d.f.	Tail probability
PDR	1454.3	1436				
PD/PR/DR	1457.6	1440	*PDR* interaction	3.3	4	0.51
PD/PR	1459.2	1442	*DR* interaction	1.6	2	0.45
PD/DR	1463.5	1442	*PR* interaction	5.9	2	0.052
PR/DR	1480.5	1444	*PD* interaction	22.9	4	0.0001
PD/R	1464.9	1444	*PR* interaction	5.7	2	0.058
PR/D	1481.0	1448	*PD* interaction	21.8	4	0.0002
PD	1470.7	1445	*R*	5.8	1	0.016
P/D/R	1486.0	1448	*PD* interaction	21.1	4	0.0003

Key: *P*, political affiliation; *D*, time of decision; *R*, read official leaflet.

9.3.3 Complete Stepwise

With complete stepwise, both forward selection and backward elimination are considered at each step. The process halts when the current model cannot be significantly improved by any addition, and cannot be simplified without a significant worsening of the fit. Complete stepwise will usually consider more models than the unidirectional stepwise procedures, but (with more than two variables) it will not consider every model. Comparison of Figures 9.2 and 9.3 shows that, for the referendum data, four models were not considered by either approach.

9.4 AIC AND RELATED MEASURES

In the referendum examples, the decision concerning further simplification, or further addition, to the current model was made by comparing the G^2 values of nested models. Often, however, the criterion used is the minimization or maximization of some measure of fit. For categorical data, this is usually a measure related to the Akaike Information Criterion (AIC), which is now considered.

For any data set the best-fitting model will always be the model at the top of the model tree, since this is the model containing every possible interaction between the explanatory variables. However, it will rarely be possible to explain in practical terms what such a complex model means. Indeed, such a model may be straining its sinews to explain random variations: in that case it would be best described as an *over-parameterized model* since it will contain more parameters than are really needed. What is required is a much simpler model that provides a useful working description of the data and reliable inferences concerning future observations. An example was the constant-proportion model used in Example 9.1.

A useful measure is therefore one that balances model complexity and goodness-of-fit. The most common measure is the AIC introduced by Akaike (1973, 1974). From a set of candidate models, AIC tries to select the model that provides the most adequate description of a reality that is likely to have been influenced by a myriad of unmeasured variables. The criterion has a general modeling application, but, for categorical data, it suffices to calculate

$$AIC = -2\ln(L) + 2k, \qquad (9.1)$$

where k is the number of parameters in the model and L is the maximized value of the likelihood using the model. A model with a smaller AIC value is preferred to one with a larger value.

The performance of *AIC* is improved by including a finite-sample approximation due to Sugiura (1978). The revised measure is denoted as AICc and is given by

$$AICc = -2\ln(L) + 2k\left(1 + \frac{k+1}{n-k-1}\right). \tag{9.2}$$

At present, it seems that the stepwise routines in computer packages use *AIC*. Since the difference between *AIC* and *AICc* will generally be small, the packages are unlikely to mislead the user, but it will always be worthwhile calculating the *AICc* values. Example 9.4 provides a case where the difference matters.

Very similar in spirit to *AIC* is the Bayesian Information Criterion (*BIC*) introduced by Schwarz (1978) and given by

$$BIC = -2\ln(L) + k\ln(n). \tag{9.3}$$

As with *AIC*, small values of *BIC* are preferable to large ones. Since $\ln(n)$ will be greater than 2, the optimal model chosen using *BIC* will never be more complicated than that chosen using *AIC*.

AIC and *BIC* have subtly different motivations: *AIC* seeks to select that model, from those available, that most closely resembles the true model (which will be governed by a myriad of unmeasured considerations and will *not* be amongst those considered), whereas *BIC* assumes that the correct model *is* amongst those on offer and seeks to identify that optimal model. In reality, since none of the models that we consider will ever provide a perfect description (except, perhaps, if we are studying some scientific relationship), *AICc* should be the measure used.

Example 9.3   The UK 1975 referendum (*continued*)

The four models with the lowest values of *AICc* for the UK referendum data are given in Table 9.8.

The model selected using G^2 was *PD/R* with the model *PD/PR* being rejected because the tail probability associated with the *PR* interaction was in excess of 5%. However, these *AICc* values suggest that the *PR* interaction should be seriously considered for retention.

TABLE 9.8   The values of *AICc* for the four best models for the referendum data

PD/PR	1483.4	*PD/R*	1485.1
PD/PR/DR	1485.9	*PD/DR*	1487.7

Key: *P*, political affiliation; *D*, time of decision; *R*, read official leaflet.

9.5 THE PROBLEM CAUSED BY RARE COMBINATIONS OF EVENTS

When there are many explanatory variables, each with at least two categories, some category combinations will only rarely occur. Each occurrence will correspond to either a success or a failure. For any rare combination, viewed on its own, the estimate of the probability of a success would be unreliable, because it would be based on only a small number of observations.

The iterative estimation procedure (see Section 5.4) combines information from every category combination in the process of arriving at parameter estimates. When there are rare category combinations, the likelihood function (Section 1.7) may look more like Table Mountain than Mount Everest, with the result that the location of its maximum is difficult to determine. The phenomenon was discussed by Hauck and Donner (1977).

This situation is usually easy to detect, provided one carefully examines the computer output, since then the uncertainty in the parameter estimates will be manifest and the associated tail probabilities will be large. Some computer programs may alert the user to the problem by reporting that there are probabilities near 0 or 1, or that there are problems with convergence of the iterative fitting process to the maximum.

Example 9.4   Fatal road accidents

The UK Department for Transport collects information on road accidents. The resulting data are freely available from the UK Data Service. The data for 2012 (set SN7431) show that, on Class A roads, under conditions that were not foggy, there were 9116 accidents classified as serious, of which 829 resulted in fatalities. Six variables that may be relevant to whether a serious accident results in a fatality are C (carriageway: dual, single), J (accident near junction: no, yes), L (speed limit: 30, 40, 50, 60+), P (precipitation: no, yes), S (surface: dry, wet, snow, ice), and W (high winds: no, yes). We will treat speed limit as a continuous variable.

With six possible explanatory variables, a reasonable starting model is the model containing all 15 two-variable interactions. This model has 9082 degrees of freedom with $AICc = 5265.66$. Stepwise selection, using minimization of AIC as the criterion, terminates with the removal of 10 of these interactions to give the "final" model as $CJ/JW/JS/LS/PS$, for which $AICc = 5248.76$. Table 9.9 provides an extract from the summary for this model.

In the extract in Table 9.9, P:S2 is contrasting the combined effect (on the probability of a fatality) of the driving conditions "precipitation and a wet road," with the alternative being "precipitation and a dry road." Note the huge

TABLE 9.9 Extract from the output (using R) for the fit of the $CJ/JW/JS/LS/PS$ model to road accident data. Variables in the table are P (precipitation: no, yes), and S (road surface: dry, wet, snow, ice)

```
       Coefficients:
                 Estimate  Std. Error  z value  Pr(>|z|)
       P:S2       1.544e+01  1.461e+03   0.011    0.9916
       P:S3      -1.228e+01  1.583e+03  -0.008    0.9938
       P:S4       1.670e+01  1.461e+03   0.011    0.9909
```

TABLE 9.10 Numbers of accidents and occurrence of fatalities under selected conditions

Road surface conditions	No precipitation				Precipitation			
	Dry	Wet	Snow	Ice	Dry	Wet	Snow	Ice
Number of accidents	6251	1384	5	126	7	1306	23	14
Number of fatalities	548	153	1	13	0	110	1	3
Percent fatal	9	11	20	10	0	8	4	21

Source: SN 7431, Road Accident Data, 2012. Reproduced with permission of the UK Data Service.

standard errors and the tail probabilities close to 1. Although "precipitation and a dry road" can occur at the start of a rainstorm, a serious accident under these conditions is not a common occurrence (as shown by the figures given in Table 9.10). The large standard errors suggest that this is not the end of the analysis (which is continued in Section 9.5.1).

9.5.1 Tackling the Problem

If some category combinations are very rare, then this may be a consequence of an entire category being rare. If that is so, then it may be advisable either to remove that category (by selecting for analysis only the data belonging to other categories) or to combine that category with some similar category. Thus, in the previous example, since there were only 28 cases where the accident occurred on snowy roads, it might be sensible to combine "Snow" and "Ice" in a single category.

Alternatively, without making any adjustments to the selected data, or to the category definitions, it may be worth proceeding by removing the imprecisely estimated interaction and examining the fit of the simplified model (since there may be other effects or interactions that are affected by the same lack of data). A stepwise procedure could then be reapplied.

Remember that a stepwise procedure stops when it cannot find a *single* group of parameters that will result in a reduction in *AIC* (or *AICc*). Situations occur not infrequently where removing *more than one* group of parameters will result in an improvement. The road accident data will illustrate this. The fact that this can occur is a reminder that it is the data analyst that is in charge of the data analysis—not the computer program!

Example 9.4   Fatal road accidents (*continued*)

The analysis of the accident data started with Model 1 (see Table 9.11), the model with all fifteen two-variable interactions. Stepwise selection ended at Model 2. However, Table 9.9 suggested that the *PS* interaction parameters were not significantly different from one another (and hence not significantly different from zero). Removing those interactions leads to Model 3, which, of course, has a higher value for *AIC* than its predecessor (since stepwise stopped

TABLE 9.11   The fit of alternative models to UK road accident data

	Description	d.f.	G^2	Change in G^2	d.f.	Tail prob.	*AICc*
1	All 15 2-variable interactions	9082	5197.3				5265.561
	Stepwise removal of ten interactions using AIC						
2	$CJ/JW/JS/LS/PS$	9096	5208.7	11.38	14	0.656	5248.764
	Removal of PS interaction because of small sample sizes (see Tables 9.9 and 9.10)						
3	$CJ/JW/JS/LS/P$	9099	5218.0	9.37	3	0.025	5252.104
	Stepwise removal of LS interaction using AIC						
4	$CJ/JW/JS/L/P$	9102	5220.7	2.71	3	0.439	5248.789
	Removal of JS interaction because of lower AICc						
5	$CJ/JW/S/L/P$	9105	5226.7	6.00	3	0.111	5248.772
	Stepwise removal of S using AIC						
6	$CJ/JW/L/P$	9108	5229.3	2.51	3	0.473	5245.272
7	$CJ/JW/L$	9109	5232.9	3.64	1	0.056	5246.908
8	$C/JW/L$	9110	5236.8	3.94	1	0.047	5248.849
9	JW/L	9111	5236.8	0.00	1	0.950	5246.850

Source: SN 7431, Road Accident Data, 2012. Reproduced with permission of the UK Data Service. Key: *C*, carriageway type; *J*, junction proximity; *W*, windy; *S*, road surface; *P*, precipitation condition; *L*, speed limit.

at Model 2). Re-applying the stepwise algorithm removes the *LS* interaction (to give Model 4), and, if *AICc* had been used, then the *JS* interaction would also be removed. This would result in the selection of Model 5. Using the stepwise algorithm once again results in a further simplification (to Model 6). Note that the values for *AICc* and *AIC* for Model 6 are less than those for Model 2. The initial stepwise algorithm was "held up," first by *PS* and subsequently by *JS*.

9.6 SIMPLICITY VERSUS ACCURACY

Ockham's razor stated that with two equally good explanations of the data, one should prefer the simpler. Measures such as *AICc* provide information concerning the appropriateness of a model, while the comparison of nested models using G^2 (as in Table 9.5) provides information about the relevance of particular effects and interactions. For data sets with many variables there will rarely be a clear-cut "best" model.

Important questions to ask are why the data were collected and why the data are being analyzed. If the data were collected to test some theory, then the data collector will have had in mind a specific model to be tested and the questions concerning model selection will scarcely exist. If what is wanted is a simple summary of past events then the simplest model may be appropriate, whereas if there is an interest in predicting the future, then a rather more complex model might be appropriate. In every case the decision should be made by a person rather than a machine.

Example 9.4   Fatal road accidents (*continued*)

After three applications of stepwise procedures, Model 6 (see Table 9.11) was selected. As it happens this is indeed the model with the smallest values of *AIC* and *AICc*. However, there are reasonable arguments for further simplification through Models 7 and 8 to Model 9. The successive changes in G^2 have magnitudes 3.64, 3.94, and 0.004. Each is to be compared with a chi-squared distribution with one degree of freedom. Since the upper 5% point of this distribution is 3.84, the first simplification just fails to be significant at that level, while the marginal significance of the second simplification is easily offset by the negligible effect (in terms of fit) of removing *C*. Of course, the elimination of *C* is anything but negligible when it comes to simplification of the description, since removal of a variable (note that road surface condition and precipitation were removed earlier) greatly helps with the model description.

To an extent, the proof of the pudding is in the eating: we need to examine the actual fit of competing models, in order to visually judge their success.

TABLE 9.12 **The fit of alternative models to UK road accident data with serious injuries. Combinations shown are those with relatively large numbers of observations (at least 200 accidents when not windy, or at least 20 accidents when windy)**

C	J	L	n	Obsd %	$CJ/JW/L/P$	JW/L
			Not windy; dry			
Dual	Junction	30 MPH	274	4.0	4.9	4.1
Single	No junction	30 MPH	939	6.7	6.3	5.9
Single	Junction	30 MPH	2528	3.8	4.0	4.1
Dual	Junction	40 MPH	205	6.8	7.2	6.0
Single	No junction	40 MPH	288	9.4	9.2	8.7
Single	Junction	40 MPH	311	7.4	5.9	6.0
Dual	No junction	60 MPH	426	16.9	16.9	17.9
Single	No junction	60 MPH	1160	17.3	18.9	17.9
Single	Junction	60 MPH	591	13.7	12.6	12.9
			Not windy; not dry			
Single	Junction	30 MPH	357	2.5	7.3	4.1
Single	No junction	60 MPH	1210	17.6	15.9	17.9
			Windy; dry			
Single	Junction	30 MPH	20	0	8.8	8.4
			Windy; not dry			
Single	Junction	30 MPH	33	12.1	7.3	8.4
Single	No junction	60 MPH	32	15.6	13.8	12.9

Table 9.12 compares the observed percentages with those estimated by Model 6 (the *AICc* selection) and Model 9 (which uses information on just three of the six original explanatory variables).

Most days are not especially windy, so most accidents happen on "not windy" days. The table gives information for category combinations involving at least 200 serious accidents under not-windy conditions, but relaxes this for windy days. For the 14 situations given in the table, Model 6 ($CJ/JW/L/P$) provides the closer estimate for 6 situations and Model 9 (JW/L) for 8 situations. Neither model is outstandingly effective, though both capture the essence of the fluctuations.

The bottom line is that fatalities are a function of speed: there is a higher proportion of fatalities on roads with a higher speed limit, and in situations away from junctions (which may cause traffic to slow).

9.7 DFBETAS

This chapter has been concerned with model selection and the difficulties of balancing the advantages of a well-fitting complex model with those of a simpler model that fits less well. Whichever model is used, we need to be convinced that it is a reasonable fit for all the data and not just for a subset of the data. Comparing observed counts or proportions with those fitted by the model, as in Table 9.12, is an essential part of model checking, but it is not the end of that checking.

When estimating the parameters of a general linear model, there are several statistics that are regularly calculated as checks on the presence of outlying or unduly influential observations. In the present context the most useful appears to be the curiously named DFBETAS. The name was introduced by Belsley, Kuh, and Welsch (1980) in the context of a model in which all the parameters were βs. For example

$$y = \beta_0 + \beta_1 x.$$

With n observations we obtain estimates $\widehat{\beta_0}$ and $\widehat{\beta_1}$. Now suppose we exclude observation i and recalculate the parameter estimates, getting the new estimates $\widehat{\beta_{0(i)}}$ and $\widehat{\beta_{1(i)}}$. These will usually be different to the original estimates. This "difference in a β value" was shortened to DFBETA.

To judge the importance of DFBETA, we work with the standardized values of the parameters (i.e., the parameter estimates divided by their standard errors). It is this standardization that accounts for the final S in DFBETAS. Belsley et al. (1980) suggested that a DFBETAS value having a magnitude greater than $2/\sqrt{n}$ indicated an *influential observation*.

Example 9.5 **Conditions affecting the growth of a bacterium in orange juice**

Table 9.13 presents the results of a series of laboratory experiments intended to identify the conditions (temperature (T), acidity (A), nisin concentration in IU/mL (N), and °Brix (B)) under which the bacterium *Alicyclobacillus acidoterrestris* grows in orange juice. Each experiment was conducted twice, with the same outcomes being observed on each occasion. There were 54 experiments in all.

With just 54 observations and four quantitative explanatory variables, a suitable starting model (see Table 9.14) might be the model with all two-variable interactions. However, using R, this elicits a warning message concerning near zero probabilities (see Section 9.5). Backward elimination using AIC stops at the model $T/AB/NB$. There is again a warning message and neither interaction appears significantly different from zero. Removing either

TABLE 9.13 Growth (1) or no growth (0) of *Alicyclobacillus acidoterrestris* CRA 7152 in orange juice under varying experimental conditions

°C	pH	Ni	°Brix	Outcomes	°C	pH	Ni	°Brix	Outcomes
28.5	3.7	17.5	13	0, 0	45.5	3.7	17.5	13	1, 1
28.5	5.1	17.5	13	1, 1	45.5	5.1	17.5	13	1, 1
28.5	3.7	52.5	13	0, 0	45.5	3.7	52.5	13	1, 1
28.5	5.1	52.5	13	0, 0	45.5	5.1	52.5	13	1, 1
28.5	3.7	17.5	17	0, 0	45.5	3.7	17.5	17	1, 1
28.5	5.1	17.5	17	1, 1	45.5	5.1	17.5	17	1, 1
28.5	3.7	52.5	17	0, 0	45.5	3.7	52.5	17	0, 0
28.5	5.1	52.5	17	0, 0	45.5	5.1	52.5	17	1, 1
20	4.4	35	15	0, 0	54	4.4	35	15	0, 0
37	3	35	15	0, 0	37	5.8	35	15	1, 1
37	4.4	0	15	1, 1	37	4.4	70	15	0, 0
37	4.4	35	11	1, 1	37	4.4	35	19	0, 0
37	4.4	35	15	1, 1	37	4.4	35	15	1, 1
37	4.4	35	15	1, 1					

Source: Peña, https://sites.google.com/a/unitru.edu.pe/sciagropecu/publicacion/scagropv1n1/scagrop01_47-61. CC BY-NC 3.0, http://creativecommons.org/licenses/bync/3.0/deed.es_ES

interaction leads to a model with the same fitted values (a side effect of the earlier warnings) and an increased AIC value. A reduction in AIC is obtained by removing the remaining interaction: the resulting model, $T/A/N/B$, is the model with the smallest values for both AIC and AICc. This data set therefore provides another example of the need to investigate models other than that suggested by an automated stepwise procedure.

In this case the suggested critical value for DFBETAS $(2/\sqrt{n})$ is 0.27; examination of the values of DFBETAS reveals one particularly influential pair of observations (see Figure 9.4) relating to the parameter measuring the effect of temperature. The reason that these particular observations are selected is because the outcome (no bacterium growth) is a sharp contrast

TABLE 9.14 The fit of alternative models to the orange juice data

Description	d.f.	G^2	Change in G^2	d.f.	Tail prob.	*AICc*	Warning
$TA/TN/TB/AN/AB/NB$	43	29.95				58.24	Yes
$T/AB/NB$	47	32.72	2.77	4	0.60	49.16	Yes
$T/A/NB$ or $T/AB/N$	48	35.48	2.76	1	0.10	49.27	No
$T/A/N/B$	49	36.32	0.84	1	0.36	47.57	No

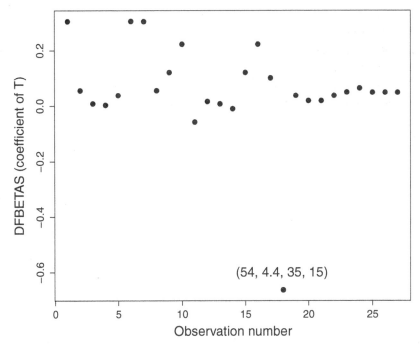

FIGURE 9.4   The values of DFBETAS for the coefficient of T for the orange juice data. The extreme value for DFBETAS corresponds to the pair of observations taken at 55°C.

to the results at the next highest temperature (seven observations of growth in eight trials). The experimental conditions deliberately included extreme values and there is no reason to suppose that this signifies any experiment error.

When there are categorical variables, there are likely to be several observations with the same values for the DFBETAS; if their common value is large then they are collectively influential. In such a case a dummy variable (Section 15.5) may lead to model simplification.

REFERENCES

Akaike, H. (1973). Information theory and an extension of the maximum likelihood principle. In: *2nd International Symposium on Information Theory*, edited by B. N. Petrov and F. Csaki, pp. 267–281, Akademia Kiado, Budapest, Hungary.

Akaike, H. (1974) A new look at the statistical model identification. *IEEE Transactions on Automatic Control*, **19** (6), 716–723.

Belsley, D. A., Kuh, E., and Welsch, R. E. (1980) *Regression Diagnostics*, John Wiley & Sons, New York.

Hauck, W. W., and Donner, A. (1977) Wald's test as applied to hypotheses in logit analysis. *J. Am. Stat. Ass.*, **72**, 851–853.

Sugiura, N. (1978). Further analysis of the data by Akaike's information criterion and the finite corrections. *Comm. Statist.* A **7**, 13–26.

CHAPTER 10

MULTINOMIAL LOGISTIC REGRESSION

This chapter extends the ideas of Chapters 7–9, where the response variable had just two categories, to the case where several categories are possible. The underlying distribution is therefore now multinomial (see Section 1.6.2). Models for this situation were described as *discrete choice models* by McFadden (1974).

10.1 A SINGLE CONTINUOUS EXPLANATORY VARIABLE

In the previous chapters, the response was the logarithm of the ratio of the probabilities of the two categories of a binary variable. Now, with several categories, there are many possible probability ratios that could be examined.

With a response variable having J categories, computer programs usually choose one category (normally either category 1 or category J) as the single reference category. A typical multinomial logistic regression model with a single continuous explanatory variable (x) is

$$\ln(p_j/p_1) = \mu_j + \alpha_j x, \qquad j = 2, 3, \ldots, J, \qquad (10.1)$$

where p_j is the probability of occurrence of the jth category of the response variable. This model consists of $(J - 1)$ simultaneous equations with $2(J - 1)$

Categorical Data Analysis by Example, First Edition. Graham J. G. Upton.
© 2017 John Wiley & Sons, Inc. Published 2017 by John Wiley & Sons, Inc.

parameters to be estimated. Notice that the equations are simultaneous, not separate; computer programs will estimate all the parameters simultaneously.

Define E_j by

$$E_j = p_j/p_1 = \exp(\mu_j + \alpha_j x), \qquad \text{for } j = 2, 3, \ldots, J, \quad (10.2)$$

with $E_1 = 1$. Thus

$$p_j = E_j p_1, \qquad j = 1, \ldots, J. \qquad (10.3)$$

Since $p_1 + p_2 + \cdots + p_J = 1$,

$$\left(\sum_{j=1}^{J} E_j \right) p_1 = 1,$$

and hence

$$p_j = E_j \bigg/ \sum_{j=1}^{J} E_j. \qquad (10.4)$$

Example 10.1   The dependence of political allegiance on age

In early February 2007, members of a YouGov panel were interviewed and asked, amongst other questions, to state their age and political allegiance. There were 2890 respondents, with 1287 expressing no particular allegiance and 141 expressing allegiance to minor parties. Table 10.1 reports the results of the respondents who expressed allegiance to one of the principal political parties.

TABLE 10.1   Variations in percentages of voters stating an allegiance to a particular political party, subdivided by age

Political allegiance	Age range					
	< 25	25–34	35–44	45–54	55–64	≥ 65
Labour	45	51	54	47	45	38
Conservative	42	39	37	41	44	55
Liberal Democrat	13	10	9	12	11	8

Source: SN 6322, Gender and the Vote in Britain, 2007 Reproduced with permission of the UK Data Service.

The table suggests that there is no uniform age-related trend, though it is the case that the final four age categories show a steadily increasing proportion of Conservative supporters.

Whilst Equation (10.1) will not provide a good description for the entire age range, the table suggests that it might do a reasonable job of describing the allegiances of those aged 35 and above. Note that the original data are recorded as age in years; the grouping in the table is only to examine for underlying trends. Model fitting uses the ungrouped data. The model given by Equation (10.1) is fitted using the following R commands:

R code

```
# age and support are vectors
w<-which((age>34));
summary(multinom(support[w]~age[w]))
```

An extract from the resulting output is shown in Table 10.2.

TABLE 10.2   Extract from the output (using R) for the fit of Equation (10.1) to the party allegiance data of those aged 35 and above. The values shown are the estimates of the α-parameters and their standard errors

```
Coefficients:
                   (Intercept)         Age
Conservative         -1.333971  0.023344509
Liberal Democrat     -1.806401  0.005382087
Std. Errors:
                   (Intercept)         Age
Conservative          0.3355707  0.005907915
Liberal Democrat      0.5333409  0.009524643
```

If a respondent's age was unrelated to that person's political affiliation, then the estimates of the α-parameters would not differ significantly from zero. To examine their significance we divide each estimate by the corresponding standard error. The first row 'Conservative' refers to the logit comparing Conservative allegiance with Labour allegiance. The ratio of the estimate to its standard error (0.0233 … /0.0059 …) is 3.95, which is much larger than the standard 5% point (1.96). Since the estimate is positive, this indicates that allegiance to the Conservative party (as opposed to the Labour party) significantly increases with age. The corresponding ratio for the Liberal Democrats is also positive, but, at 0.57, it is not significant.

To get a clearer picture, we choose two specific ages (40 and 60) and use the results from Table 10.2 and Equation (10.4) to obtain the specific estimated allegiance percentages for those ages. The calculations are summarized in Table 10.3.

TABLE 10.3   Calculations of estimated party-allegiance percentages for respondents aged 40 and 60, using Equations (10.2)-(10.4) and the parameter estimates reported in Table 10.2

Age (x)	E_{Lab}	E_{Con}	E_{Lib}	$\sum E_j$	p_{Lab}	p_{Con}	p_{Lib}
40	1	0.67	0.20	1.87	0.53	0.36	0.11
60	1	1.07	0.23	2.30	0.44	0.47	0.10

Note that, although α_{Lib} is positive (= 0.0053 ...), indicating an improved performance relative to the Labour party, this does not correspond to a rise in Liberal Democrat support. This is due to the much steeper rise in support for

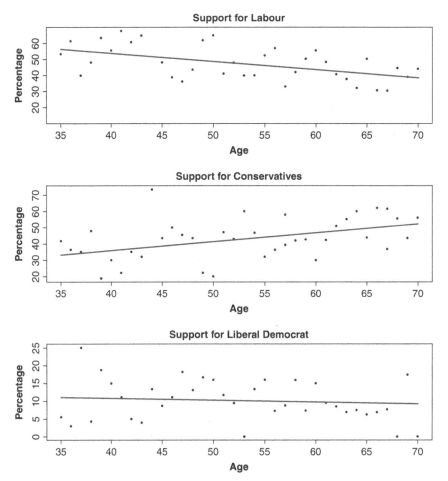

FIGURE 10.1   The fit of the model describing the dependence of political allegiance on age.

the Conservative party. The estimates, which were derived from simultaneous equations, must also be interpreted on a simultaneous basis.

In Figure 10.1, the fitted lines appear to be straight. In reality each is a small section from the center of a logit curve (see Figure 7.1).

10.2 NOMINAL CATEGORICAL EXPLANATORY VARIABLES

Recall that a variable is described as being *categorical* if the 'values' that it takes are neither continuous (e.g., height) nor discrete (e.g., number of persons in a household). If the categories have no special order (e.g., types of fruit) then the variable is a *nominal categorical variable*, whereas if the categories are ordered (e.g., "in favor," "neutral," "against") then the variable is an *ordinal categorical variable*.

Unless there are many category combinations, data involving only categorical variables can be easily summarized using a contingency table. To understand the effects (if any) of the explanatory variables on the response variable, it may be helpful to report the data as proportions for the categories of the response variable.

Example 10.2   The dependence of political allegiance on gender and social class

The data analyzed in the previous example were collected with the aim of investigating the relevance of gender on political affiliation. Since it is well known that social class is an important determinant of an individual's politics, we now include that as a second categorical explanatory variable. The data are reported in Table 10.4.

TABLE 10.4   Percentages supporting the three principal political parties, with subdivisions by gender and social class

| | Social classes A, B, and C1 | | Social classes C2, D, and E | |
	Male	Female	Male	Female
Sample size	446	471	333	250
Labour	42.2	41.6	55.0	51.2
Conservative	47.5	47.3	33.3	41.6
Liberal Democrat	10.3	11.0	11.7	7.2

Source: SN 6322, Gender and the Vote in Britain, 2007. Reproduced with permission of the UK Data Service.

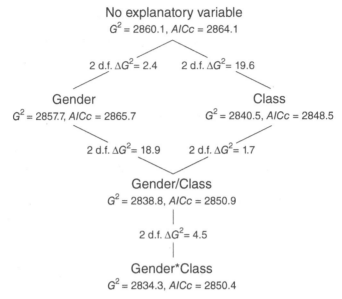

FIGURE 10.2 The tree of models describing the dependence of political allegiance on gender and/or class.

The table shows little difference between the political affiliations of males and females in the upper social classes (A, B, and C1), but in the lower social classes, where there is a strong preference for the Labour party, this preference is a little greater for males than females. This is, in part, a consequence of an apparent relative dislike for the Liberal Democrat party amongst females from the lower social classes. In summary, therefore, before we undertake a formal analysis, we can conclude that political affiliation is certainly related to class, and that there may be a gender×class interaction.

Figure 10.2 reports the goodness of fit of the five possible models, and also the differences in fit for nested pairs of models. Since the response variable has 3 categories and each explanatory variable has 2 categories, there are $(3 - 1) * (2 - 1) = 2$ d.f. for the difference between every adjacent pair of models.

The model with the lowest value for $AICc$ (see Section 9.4) is the model that takes account only of class. There is no significance attached to the introduction of gender either as the first explanatory variable ($\Delta G^2 = 2.4$; 2 d.f.) or to the model that includes class ($\Delta G^2 = 1.7$; 2 d.f.). However, as the reduction in $AICc$ suggested, the improvement in fit ($\Delta G^2 = 19.6$; 2 d.f.) due to including class in the model is highly significant ($p \approx 2 \times 10^{-7}$).

While the change in G^2 resulting from the inclusion of the gender-class interaction is not significant ($\Delta G^2 = 4.5$; 2 d.f.; p = 0.20), the model that

includes the interaction does have the second lowest value for *AICc*, implying that it was certainly worth investigating whether the interaction was
required.

10.3 MODELS FOR AN ORDINAL RESPONSE VARIABLE

When the categories of the response variable are ordered, there are many possible comparisons that take account of that ordering and many different types
of model have been proposed. A useful survey is provided by Liu and Agresti
(2005) while entire books (e.g., Clogg and Shihadeh, 1994; Agresti, A., 2010)
have been devoted to the topic. This section therefore provides no more than
an introduction.

10.3.1 Cumulative Logits

Suppose that there are J ordered categories with the probability of an individual belonging to category j being p_j. For convenience, assume that the "lowest" category is category 1, and the "highest" is category J. The categories
are ordered, but not necessarily numerical: category 1 might be "strongly in
favor" and category J might be "strongly against" (or vice versa). Define P_j
to be the cumulative probability of an individual belonging to category j or
lower, so that

$$P_j = p_1 + \cdots + p_j, \qquad j = 1, 2, \ldots, J - 1. \qquad (10.5)$$

The *cumulative logit* L_j is defined by

$$L_j = \ln\left(\frac{P_j}{1 - P_j}\right) = \ln\left(\sum_{k=1}^{j} p_k \Big/ \sum_{k=j+1}^{J} p_k\right), \quad j = 1, 2, \ldots, J - 1. \quad (10.6)$$

For example, if the observed counts are

$$10, \qquad 30, \qquad 20,$$

then the observed values for the cumulative logits are

$$L_1 = \ln\left(\frac{10}{30 + 20}\right) = \ln(0.2) = -1.61,$$
$$L_2 = \ln\left(\frac{10 + 30}{20}\right) = \ln(2) = 0.69.$$

The simplest model, in which the response variable is unaffected by explanatory variables, is

$$L_j = \mu_j, \qquad j = 1, 2, \ldots, J - 1. \tag{10.7}$$

Our discrete response variable has ordered categories. Although these categories may be described using words (e.g., "strongly agree"), it is not difficult to imagine that there is an (unmeasured) underlying variable (a so-called *latent variable*) that measures agreement on a continuous scale. The $\{\mu_j\}$, which are often referred to as *cutpoint parameters*, correspond to those values of the latent continuous variable where the observed response variable changes from one category to the next. Note that $\mu_1 \leq \mu_2 \leq, \ldots, \leq \mu_J$.

10.3.2 Proportional Odds Models

These models may involve any number of explanatory variables, but, for simplicity, we limit discussion to cases involving a single explanatory variable, X.

When X is continuous the proportional odds model is:

$$L_j(x) = \mu_j + \beta x, \qquad j = 1, 2, \ldots, J - 1. \tag{10.8}$$

With this model, the difference between cumulative logits for two different values of X (x_1 and x_2, say) is linearly dependent on the difference between the values of X, and is the same for all j:

$$\ln \left(\frac{\sum_{k=1}^{j}(p_k|X = x_1) / \sum_{k=j+1}^{J}(p_k|X = x_1)}{\sum_{k=1}^{j}(p_k|X = x_2) / \sum_{k=j+1}^{J}(p_k|X = x_2)} \right) = L_j(x_1) - L_j(x_2)$$

$$= (\mu_j + \beta x_1) - (\mu_j + \beta x_2)$$

$$= \beta(x_1 - x_2).$$

The argument of this logarithm is a *cumulative odds-ratio*. The fact that the magnitude of the logarithm is proportional to the difference between the values of X explains the description as a proportional odds model (McCullagh, 1980).

Table 10.5 shows examples of the cell probabilities resulting from Equation (10.8) with three categories, $\beta = 1$, $\mu_1 = 0.2$, and $\mu_2 = 2$. Figure 10.3 illustrates the same situation for values of x between -5 and 5.

If instead X is categorical with I categories, then a simple model is

$$L_j(x_i) = \mu_j + \beta_i, \qquad i = 1, 2, \ldots, I; j = 1, 2, \ldots, J - 1. \tag{10.9}$$

TABLE 10.5   Category probabilities resulting from the cumulative-logit model given by Equation (10.8) for the case $\beta = 1$

x	-1	0	1
$\mu_1 = 0.2$	$-2.609 = \ln(0.0736)$	$\ln(0.2)$	$-0.609 = \ln(0.544)$
$\mu_2 = 2$	$-0.307 = \ln(0.736)$	$\ln(2)$	$1.693 = \ln(5.44)$
(p_1, p_2, p_3)	$(0.07, 0.35, 0.58)$	$(0.17, 0.50, 0.34)$	$(0.35, 0.49, 0.16)$

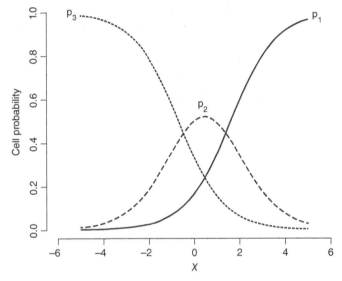

FIGURE 10.3   The dependence of the cell probabilities on the value of x, using the cumulative logit model given by Equation (10.8) with three categories, $\beta = 1$, $\mu_1 = 0.2$, and $\mu_2 = 2$.

Table 10.6 provides an example of data that provide a perfect fit to this model. The frequencies for x_1 are those given previously. Although the pattern of the frequencies for x_2 appears entirely unrelated to those for x_1, this is not the case, since the cumulative logits for x_2 are

$$\ln\left(\frac{48}{18 + 22}\right) = \ln(1.2) \text{ and } \ln\left(\frac{48 + 18}{22}\right) = \ln(3).$$

TABLE 10.6   A 2 × 3 table exactly satisfying the cumulative-logit model given by Equation (10.9)

j	1	2	3
x_1	10	30	20
x_2	48	18	22

Comparison with the corresponding values for x_1 shows that the data satisfy $\mu_1 = 0, \mu_2 = 1, \beta_1 = 0.2,$ and $\beta_2 = 2$.

Example 10.3   Working hours in Great Britain in 2015

The Labour Force Survey is a quarterly sample survey of approximately 56,000 households living at private addresses in Great Britain. Table 10.7 summarizes the hours of work (excluding overtime) reported by those surveyed in the second quarter of 2015.

The distributions of the hours of work vary markedly across the four rows with females working on average fewer hours than their male counterparts. No doubt they are fulfilling family commitments; a respondent's age and family condition would be extra explanatory variables that would need to be included to gain a proper understanding of the data.

In this case the two explanatory variables are marriage (x) and sex (y), each at two levels. Using cornered constraints (see Section 11.2), Equation (10.9) can be rewritten as

$$L_j(x, y) = \mu_j + \beta x + \gamma y + \delta xy, \qquad j = 1, 2, \ldots, J-1; x = 0, 1; y = 0, 1.$$

$$(10.10)$$

In this formulation β measures the marriage effect, γ measures the sex effect, and δ measures the interaction effect.

Specialized computer routines are available to fit cumulative logit models. The R code follows

TABLE 10.7   The hours of work (excluding overtime) reported by Great Britain's Labour Force Survey in the second quarter of 2015

Married?	Sex	Hours of work						
		≤ 19	20–29	30–34	35–39	40	≥ 41	Total
Yes	Male	313	368	513	2479	1374	853	5900
	Female	880	941	558	1281	361	199	4220
No	Male	215	145	190	772	421	223	1966
	Female	429	380	273	783	228	112	2205

Source: SN 7725, Quarterly Labour Force Survey, April–June, 2015. Reproduced with permission of the UK Data Service. Married includes co-habiting or living with a civil partner.

R code

```
library(VGAM);
hrs<-read.table("hours.dat",header=TRUE);
model<-vglm(cbind(h0to19, h20to29, h30to34,
    h35to39, h40, h41up) ~sex*married,
family=cumulative(parallel=TRUE),data=hrs);
summary(model)
```

An extract from the resulting output is shown in Table 10.8.

The first thing to notice in the output is that the Pearson residuals (Equation 4.6) refer to the cumulative logits rather than the cell frequencies because it is their values that are being modeled. The model is using five cutpoint parameters and three parameters that measure the effects of the explanatory variables. There are 20 cumulative logits and 8 parameters, so there are 12 degrees of freedom.

The next thing to notice is that, with cornered constraints, since R is reporting a value for sexM:marriedY the reference categories are the last ones encountered: "Female" and "Not married." We can check this by calculating the observed value of the cumulative logits for the unmarried females.

TABLE 10.8 **Extract from the output (using R) for the fit of Equation (10.10) to the working hours data of Table 10.7**

```
Pearson Residuals:
L(P[Y<=1])    L(P[Y<=2])   L(P[Y<=3])   L(P[Y<=4])    L(P[Y<=5])
1 -0.54743    -4.71350     -0.6828772    2.12617       1.43532
2 -3.03868     3.91519      1.3320121   -1.60072      -3.09223
3  5.68272    -1.43623     -0.0014623   -2.02470       0.36156
4  1.48378     0.91776     -1.2294938   -0.54176      -0.77294


Coefficients:
                Estimate Std. Error   z value
(Intercept):1 -1.494242   0.042229 -35.38419
   ...           ...         ...        ...
(Intercept):5  3.010607   0.047757  63.03953
sexM          -0.958641   0.056337 -17.01604
marriedY       0.228857   0.046937   4.87584
sexM:marriedY -0.559319   0.066593  -8.39909


Residual deviance: 110.1943 on 12 degrees of freedom
```

Thus

$$\ln\left(\frac{429}{2205-429}\right) = -1.42, \text{ and } \ln\left(\frac{2205-112}{112}\right) = 2.93,$$

are reassuringly close to the model estimates (to 2 d.p.) of -1.49 and 3.01. Different routines and different programming languages may use alternative constraints. It is always wise to check that the output makes sense.

Since several of the residuals have magnitudes greater than three, and the chance of a chi-squared distribution with twelve degrees of freedom taking a value greater than 110 (the residual deviance reported in the output) is negligible, it is clear that the model does not provide a statistically acceptable fit.

To better understand the lack of fit of the model, the following R code expresses the fitted values obtained using the model as percentages of the values observed. The results are shown in Table 10.9. The model certainly underestimates the proportion of unmarried men working small numbers of hours, but, for the bulk of the data, it appears to perform adequately.

R code (*continued*)

```
round(100*fitted.values(model)*rowSums(hours)/hours,0)
```

Another way of assessing the goodness of fit of a model is to compare how well it does against alternative models. Figure 10.4 summarizes the fit of alternative proportional odds models for these data. Although the interaction between the two explanatory variables is certainly not negligible (change in $G^2 = 68.8$, 1 d.f.), it is obvious that the most important effect in Table 10.7 is the difference in the working hours of males and females (change in $G^2 >$ 1800, 1 d.f.).

TABLE 10.9   The fitted values obtained using Equation (10.10) expressed as percentages of the values observed

Married?	Sex	Hours of work					
		≤ 19	20–29	30–34	35–39	40	≥ 41
Yes	Male	110	**122**	92	95	101	105
	Female	106	**87**	106	107	98	**80**
No	Male	**72**	**133**	101	105	93	100
	Female	94	102	**111**	100	98	92

Bold type indicates errors of more than 10%.

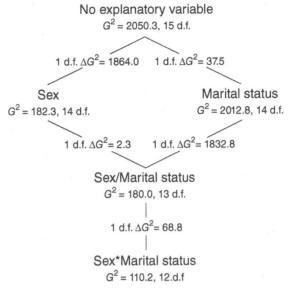

No explanatory variable
$G^2 = 2050.3$, 15 d.f.

1 d.f. $\Delta G^2 = 1864.0$ 1 d.f. $\Delta G^2 = 37.5$

Sex
$G^2 = 182.3$, 14 d.f.

Marital status
$G^2 = 2012.8$, 14 d.f.

1 d.f. $\Delta G^2 = 2.3$ 1 d.f. $\Delta G^2 = 1832.8$

Sex/Marital status
$G^2 = 180.0$, 13 d.f.

1 d.f. $\Delta G^2 = 68.8$

Sex*Marital status
$G^2 = 110.2$, 12.d.f

FIGURE 10.4   The tree of models describing the dependence of working hours on sex and marital condition using proportional odds models.

For an approach providing a formal assessment of the fit of a proportional odds model, see Brant (1990).

10.3.3 Adjacent-Category Logit Models

Instead of comparing each category with (for example) category 1, in these models each category is compared with its predecessor. Since

$$\ln(p_{j+1}/p_j) = \ln(p_{j+1}/p_1) - \ln(p_j/p_1),$$

any adjacent-category logit model is really no more than a re-parameterization of a multinomial logistic regression model.

Example 10.3   Working hours in Great Britain in 2015 (*continued*)

Continuing with the previous example it is again possible to take advantage of routines developed by others:

R code (*continued*)

```
model<-vglm(cbind(h0to19, h20to29, h30to34, h35to39,
    h40, h41up)~sex*married,
    family=acat(parallel=TRUE),data=hours)
```

The model has as many parameters as the previous proportional odds model, but the fit is slightly worse ($G^2 = 120.8$ as opposed to 110.1).

10.3.4 Continuation-Ratio Logit Models

Cumulative logits involved all J categories of the ordinal response variable. Adjacent-category models used just two categories. By contrast, continuation-ratio logit models use variable numbers of categories, dependent on the value of j. These models address the question of whether an observation belongs to category j, or to a more extreme category. There are two possible sets of logits, depending on whether "extreme" means above or below. For the "above" case we have

$$L_j^A = \ln \left(\frac{p_{j+1} + \cdots + p_J}{p_j} \right), \qquad j = 1, 2, \ldots, J - 1,$$

giving successive continuation-ratio logits as

$$\ln \left(\frac{p_2 + p_3 + \cdots + p_J}{p_1} \right), \ln \left(\frac{p_3 + p_4 + \cdots + p_J}{p_2} \right), \ldots, \ln \left(\frac{p_J}{p_{J-1}} \right).$$

For the "below" case we have:

$$L_j^B = \ln \left(\frac{p_1 + \cdots + p_j}{p_{j+1}} \right), \qquad j = 1, 2, \ldots, J - 1,$$

giving successive continuation-ratio logits as

$$\ln \left(\frac{p_1}{p_2} \right), \ln \left(\frac{p_1 + p_2}{p_3} \right), \ldots, \ln \left(\frac{p_1 + p_2 + \cdots + p_{J-1}}{p_J} \right).$$

With a single categorical explanatory variable, X with I categories, a typical model would be

$$L_j^A(x_i) = \mu_j + \beta_i, \qquad i = 1, 2, \ldots, I; j = 1, 2, \ldots, J - 1. \quad (10.11)$$

Example 10.4   Optimism across Europe

The European Quality of Life Survey is carried out every 4 years across more than 30 European countries. Table 10.10 shows the extent to which the respondents in 2011 agreed or disagreed with the statement "I am optimistic about the future." The table suggests that the Finns were generally optimistic about

TABLE 10.10   The extent, in 2011, to which respondents agreed or disagreed with the statement "I am optimistic about the future"

Country	Strongly agree	Agree	Neither agree nor disagree	Disagree	Strongly disagree	Total
Finland	255	506	160	90	8	1019
France	236	656	494	622	260	2268
Spain	246	611	310	296	46	1509
UK	244	921	527	444	94	2230

Source: European Quality of Life Time Series, 2007 and 2011: Open Access, a part of SN 7348 the European Quality of Life integrated data file. Reproduced with permission of the UK Data Service.

the future, whereas the French were much more pessimistic. It is certainly clear that, in 2011, the degree of optimism about the future varied considerably between countries.

The following code fits the model given by Equation (10.11) with results summarized (in part) in Table 10.11.

R code (*continued*)

```
#SA is the vector c(255,236,246,244) etc
model<-vglm(cbind(SA,A,N,D,SD)~country,
     family=cratio(parallel=TRUE),data=Opt)
```

Examining the output, we see that Finland is the reference country (since it is not mentioned). As a check, the observed value of L_1^A for Finland is $\ln((1019 - 255)/255) = 1.10$ reassuringly close to the estimate for μ_1 of 1.08.

TABLE 10.11   Extract from the output (using R) for the fit of Equation (10.11) to the optimism data of Table 10.10

```
Coefficients:
              Estimate Std. Error z value
(Intercept):1  1.08033   0.053188   20.312
(Intercept):2 -0.59901   0.053085  -11.284
(Intercept):3 -0.67456   0.060040  -11.235
(Intercept):4 -2.25539   0.076638  -29.429
countryFrance  1.27716   0.058111   21.978
countrySpain   0.64220   0.061163   10.500
countryUK      0.78911   0.057464   13.732

Residual deviance: 38.47334 on 9 degrees of freedom
```

The individuals from the other countries were (as anticipated) significantly more pessimistic; markedly so in the case of the French respondents.

Although the model cannot be described as a good fit, since $P(\chi_9^2 > 38.47 \approx 0.00001)$, it is a very much better fit than the equivalent model using L^B ($G^2 = 155.4$), or the adjacent-category model ($G^2 = 62.7$), or the proportional odds model ($G^2 = 87.8$).

REFERENCES

Agresti, A. (2010) *Analysis of Ordinal Categorical Data*, 2nd ed., John Wiley & Sons, Inc., Hoboken, NJ.

Brant, R. (1990) Assessing proportionality in the proportional odds model for ordinal logistic regression. *Biometrics*, **46**, 1171–1178.

Clogg, C. C., and Shihadeh, E. S. (1994) *Statistical Models for Ordinal Variables*, Sage, Beverly Hills, CA.

Liu, I., and Agresti, A. (2005) The analysis of ordered categorical data: an overview and a survey of recent developments. *Test*, **14**, 1–73.

McCullagh, P. (1980) Regression models for ordinal data (with discussion). *J. R. Stat. Soc. B*, **42**, 109–142.

McFadden, D. L. (1974) Conditional logit analysis of qualitative choice behaviour. In *Frontiers in Econometrics*, edited by P. Zarembka, Academic Press, New York.

CHAPTER 11

LOG-LINEAR MODELS FOR
$I \times J$ TABLES

Chapters 7–9 focused on the situation where there is a single response variable with two categories. The case where the single response variable has more than two categories was discussed in Chapter 10. This chapter introduces log-linear models, which are particularly appropriate when there is more than one response variable. A log-linear model can also be used with a single response variable, though in this case precisely the same results will be obtainable using a simpler looking logistic model.

11.1 THE SATURATED MODEL

The term *saturated model* refers to any model that includes as many parameters as there are "observations" in need of explanation. For an $I \times J$ table the observations are the IJ cell frequencies.

The saturated model for an $I \times J$ table, with variables denoted by A (categories $i = 1, 2, \dots, I$) and B (categories $j = 1, 2, \dots, J$) is

$$v_{ij} = \ln(Np_{ij}) = \mu + \lambda_i^A + \lambda_j^B + \lambda_{ij}^{AB}, \qquad (11.1)$$

where p_{ij} is the probability of one of the N observations belonging to cell (i, j).

Constraints are required to reduce the number of independent parameters to a maximum equal to the observed number of cell frequencies (in this case

Categorical Data Analysis by Example, First Edition. Graham J. G. Upton.
© 2017 John Wiley & Sons, Inc. Published 2017 by John Wiley & Sons, Inc.

to IJ). Any constraints could be used, but there are two sets of constraints typically found in the literature and used in computer routines. These are discussed in Subsections 11.1.1 and 11.1.2 which contain rather a lot of (fortunately simple!) equations.

After application of either set of constraints the numbers of parameters (for an $I \times J$ table) will be as follows:

Parameter	Function	Number
μ	Measure of typical cell frequency	1
λ_i^A	Measures the relative probability of category i of variable A	$(I - 1)$
λ_j^B	Measures the relative probability of category j of variable B	$(J - 1)$
λ_{ij}^{AB}	Measures the relative probability of cell (i, j) compared to the value that would have occurred if variables A and B had been independent of one another	$(I - 1)(J - 1)$
Total		IJ

11.1.1 Cornered Constraints

Cornered constraints are often the default choice for computer programs (because they simplify the programmer's task). Here each subsequent category of a variable is compared with the first (or last) category encountered (the effect of which is set to zero). Equations (7.8) and (7.9) were examples of cornered constraints used in the context of logistic regression. In the present context, for an $I \times J$ table, with the first category chosen as the reference category, the constraints are as follows:

$$\lambda_1^A = 0, \quad \lambda_1^B = 0, \quad \lambda_{i1}^{AB} = 0 \text{ for } i = 2, \dots, I, \quad \lambda_{1j}^{AB} = 0 \text{ for } j = 2, \dots, J. \tag{11.2}$$

Applying these constraints gives:

$$v_{11} = \mu, \tag{11.3}$$

$$v_{i1} = \mu + \lambda_i^A, \qquad\qquad\qquad i \neq 1, \tag{11.4}$$

$$v_{1j} = \mu + \lambda_j^B, \qquad\qquad\qquad j \neq 1, \tag{11.5}$$

$$v_{ij} = \mu + \lambda_i^A + \lambda_j^B + \lambda_{ij}^{AB}, \qquad i \neq 1, j \neq 1. \tag{11.6}$$

Substitution of Equation (11.3) into Equations (11.4) and (11.5) gives

$$\lambda_i^A = v_{i1} - \mu = v_{i1} - v_{11} = \ln(p_{i1}/p_{11}), \tag{11.7}$$
$$\lambda_j^B = v_{1j} - \mu = v_{1j} - v_{11} = \ln(p_{1j}/p_{11}). \tag{11.8}$$

Finally, substitution of Equations (11.3), (11.7), and (11.8) into Equation (11.6) gives

$$\lambda_{ij}^{AB} = v_{ij} - v_{11} - (v_{i1} - v_{11}) - (v_{1j} - v_{11})$$
$$= v_{ij} - v_{1j} - v_{i1} + v_{11} = \ln(p_{ij}p_{11}/p_{i1}p_{1j}). \tag{11.9}$$

We see that the "main-effect" parameters λ_i^A and λ_j^B are logarithms of odds. Furthermore, since

$$p_{ij}p_{11}/p_{i1}p_{1j} = \frac{p_{11}/p_{ij}}{p_{i1}/p_{ij}},$$

the two-variable interaction parameter λ_{ij}^{AB} is the logarithm of an odds-ratio. Odds and odds-ratios were introduced in Section 3.5.3.

In a similar way it can be shown that the λ-parameters for 3-variable interactions can be interpreted as the logarithms of ratios of odds-ratios (and likewise for more complex interactions).

Example 11.1   Belief in the existence of God

The 1970 British Cohort Study (BCS70) follows the lives of people born in a single week of 1970. In 2012, the panel members were asked about the extent to which they believed in God. Since there were six possible replies to the question, a multinomial logistic regression model would be appropriate. For present purposes, however, the data are summarized in Table 11.1 using just two categories.

TABLE 11.1   Replies given by members of the 1970 British Cohort Study when interviewed in 2012 concerning their belief in God

	Non-believers	Believers
Male	2718	1321
Female	2181	2359

Source: SN 7473, 1970 British Cohort Study: Forty-Two-Year Follow-Up, 2012. Reproduced with permission of the UK Data Service.

Using R, to obtain the parameter estimates for the saturated model the following commands can be used:

R code

```
god<-data.frame(expand.grid(A=c("Male","Female"),
    B=c("No","Yes")),counts=c(2718,1321,2181,2359));
glm(counts~A*B,data=god,family="poisson")$coef
```

Since the default for R's `glm` command is the use of cornered constraints, the output is that in Table 11.2.

TABLE 11.2   Output (using R) showing the parameter estimates for the data in Table 11.1 using cornered constraints

(Intercept)	AFemale	BYes	AFemale:BYes
7.9076516	-0.7215073	-0.2201128	0.7999616

Using cornered constraints, the first category combination (Males not believing) has been taken as the reference category. This is apparent because the output refers to the remaining variable categories and not to males nor to those not believing. The "Males not believing" category is therefore being used as the baseline for comparisons and is described as "(Intercept)" in the output. Since the saturated model is an exact fit to the data, it follows that the number of male non-believers is

$$\exp(\mu) = \exp(7.9076516) = 2718.$$

The count for female non-believers requires the inclusion of the second category of the gender parameter:

$$\exp\left(\mu + \lambda_2^A\right) = \exp(7.9076516 - 0.7215073) = 1321.$$

Similarly, the count for male believers requires the inclusion of the second category of the gender parameter:

$$\exp\left(\mu + \lambda_2^B\right) = \exp(7.9076516 - 0.2201128) = 2181.$$

Finally, the count for female believers requires the addition of all the parameter types:

$$\exp\left(\mu + \lambda_2^A + \lambda_2^B + \lambda_{22}^{AB}\right) = \exp(7.9076516 - 0.7215073 - 0.2201128 + 0.7999616) = 2359.$$

11.1.2 Centered Constraints

At a time when no computer was available, centered constraints were used because they simplified the calculations in the analysis of balanced experimental designs.

Centered constraints pay equal attention to all categories of a variable and all cells in the table and may therefore be preferred to cornered constraints if the latter appear to pay undue attention to a category of no special interest. Equation (7.10) provided an example of centered constraints in the context of logistic regression. For an $I \times J$ table the constraints are as follows:

$$\sum_{i=1}^{I} \lambda_i^A = 0, \quad \sum_{j=1}^{J} \lambda_j^B = 0, \quad \sum_{i=1}^{I} \lambda_{ij}^{AB} = 0 \text{ for all } j, \quad \sum_{j=1}^{J} \lambda_{ij}^{AB} = 0 \text{ for all } i.$$

$$(11.10)$$

The nature of the parameters μ, λ_i^A, and λ_j^B is revealed by summation:

$$v_{00} = \sum_i \sum_j v_{ij} = IJ\mu, \tag{11.11}$$

$$v_{0j} = \sum_i v_{ij} = I\mu + I\lambda_j^B, \tag{11.12}$$

$$v_{i0} = \sum_j v_{ij} = J\mu + J\lambda_i^A. \tag{11.13}$$

From Equation (11.11)

$$\mu = v_{00}/IJ. \tag{11.14}$$

Substitution of Equation (11.14) into Equations (11.12) and (11.13) gives

$$\lambda_i^A = v_{i0}/J - v_{00}/IJ, \tag{11.15}$$
$$\lambda_j^B = v_{0j}/I - v_{00}/IJ. \tag{11.16}$$

Finally, substitution of these results gives

$$\lambda_{ij}^{AB} = v_{ij} - v_{00}/IJ - (v_{i0}/J - v_{00}/IJ) - (v_{0j}/I - v_{00}/IJ),$$

which simplifies to

$$\lambda_{ij}^{AB} = v_{ij} - v_{i0}/J - v_{0j}/I + v_{00}/IJ. \tag{11.17}$$

Although the algebra somewhat obscures matters, with centered constraints μ is the logarithm of the product of the total number of observations and the geometric mean of the cell probabilities; it therefore corresponds to an "average" cell frequency.

The parameter λ_i^A compares the probability of an individual being in category i of variable A with the probability for an "average" category of A. A positive value for λ_i^A implies that category i is relatively common, whereas a negative value corresponds to a relatively uncommon category.

A positive value for λ_{ij}^{AB} indicates a cell with a frequency greater than would be expected if A and B had been independent.

Example 11.1   Belief in the existence of God (*continued*)

Using R's glm command with centered constraints requires the following:

R code (*continued*)

```
contrasts(god$A)<-contr.sum(2);
contrasts(god$B)<-contr.sum(2);
glm(counts~A*B,data=god,contrasts=TRUE,
family="poisson")$coef
```

The resulting output is given in Table 11.3.

TABLE 11.3   Output (using R) showing the parameter estimates for the data in Table 11.1 using centered constraints

(Intercept)	A1	B1	A1:B1
7.63683194	0.16076324	-0.08993399	0.19999040

Using the parameter estimates given in Table 11.3, we get:

$$\exp(v_{11}) = \exp(7.63683194 + 0.16076324$$
$$+ (-0.08993399) + 0.19999040) = 2718,$$
$$\exp(v_{12}) = \exp(7.63683194 + 0.16076324$$
$$- (-0.08993399) - 0.19999040) = 2181,$$
$$\exp(v_{21}) = \exp(7.63683194 - 0.16076324$$
$$+ (-0.08993399) - 0.19999040) = 1321,$$
$$\exp(v_{22}) = \exp(7.63683194 - 0.16076324$$
$$- (-0.08993399) + 0.19999040) = 2359.$$

Because this is the saturated model, the cell frequencies given by the estimated parameters are exactly equal to those observed.

Note that, while it makes sense to use the parameter estimates with the accuracy given (eight decimal places) for calculations, it would be sensible to use no more than three decimal places in any report. The accuracy of estimated values will be provided by their associated standard errors (not shown here).

Using R, an alternative approach employs the `loglm` command (for which centered constraints are the default). The commands are as follows:

R code (*continued*)

```
library(MASS);
Table<-table(God,Sex);
coef(loglm(~God*Sex,data=Table))
```

The resulting output, shown in Table 11.4, is not as concise as that in Table 11.3, but is possibly easier to understand, The results are, of course, the same as before.

TABLE 11.4   The output resulting from R commands using the `loglm` command to obtain parameter estimates for the data in Table 11.1

```
$'(Intercept)'
[1] 7.636832
$God             No              Yes
           0.1607632  -0.1607632
$Sex            Male           Female
          -0.08993399   0.08993399
$God.Sex              Sex
God             Male         Female
   No    0.1999904  -0.1999904
   Yes  -0.1999904   0.1999904
```

11.2 THE INDEPENDENCE MODEL FOR AN $I \times J$ TABLE

Since independence implies that there are no interactions, the independence model is the saturated model without the $(I-1)(J-1)$ interaction parameters:

$$v_{ij} = \ln(Np_{ij}) = \mu + \lambda_i^A + \lambda_j^B, \tag{11.18}$$

with $i = 1, 2, \ldots, I, j = 1, 2, \ldots, J$, and with either cornered constraints (e.g., $\lambda_1^A = 0, \lambda_1^B = 0$) or centered constraints ($\sum_i \lambda_i^A = 0, \sum_j \lambda_j^B = 0$). Since there are $(I - 1)(J - 1)$ parameters omitted from the saturated model, there are $(I - 1)(J - 1)$ degrees of freedom available to test the goodness of fit of this model. For a related justification of the number of degrees of freedom, see Section 4.2.1.

Example 11.2   Murder weapons in states of the United States

The data in Table 11.5 summarize the methods used by murderers in the six states of the United States that experienced the largest numbers of reported murders in 2013.

Figure 11.1 is a cobweb diagram (see Section 4.4.2) that suggests that there are distinct departures from independence. The most significant departure appears to be that murderers in the state of New York use relatively fewer firearms, but more knives.

Using R to test the independence model results in the output shown in Table 11.6. By default R uses cornered constraints with the first category combination being the reference combination (labelled "(Intercept)"). In this case this combination is murders in California using firearms. So we find that (allowing for round-off)

$$\exp(7.0919) = 1202.2.$$

Similarly, for murders in Texas using firearms

$$\exp(7.0919 - 0.4319) = 780.6,$$

and, for murders in Texas using knives

$$\exp(7.0919 - 0.4319 - 1.6872) = 144.4.$$

TABLE 11.5   Murder weapons according to the FBI crime report: details for the six states with the largest numbers of reported murders in 2013

State	Firearms	Knives or cutting instruments	Other weapons	Hands, fists, feet, etc.	Total
California	1224	238	191	92	1745
Texas	760	164	129	80	1133
New York	362	136	113	37	648
Michigan	440	43	106	36	625
Pennsylvania	440	52	74	28	594
Georgia	411	40	74	9	534
Total	3637	673	687	282	5279

Source: https://www.fbi.gov/about-us/cjis/ucr/crime-in-the-u.s/2013/crime-in-the-u.s.-2013/ tables/table-20/table_20_murder_by_state_types_of_weapons_2013.xls

Weapon

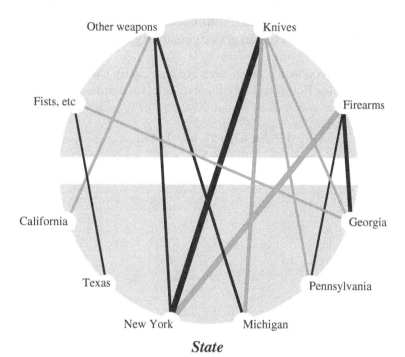

FIGURE 11.1   A cobweb diagram for the murder data of Table 11.5.

TABLE 11.6   A very abbreviated extract from the output (using R with cornered constraints) for the fit of the independence model to the murder data of Table 11.5. The full output includes parameter estimates for every state and every murder method

```
Call:  glm(formula = Freq   State + Weapon,
                 family = "poisson", data = df)
Coefficients:
        (Intercept)                 StateTexas
             7.0919                    -0.4319
     WeaponKnives   WeaponOther weapons
            -1.6872                  -1.6666
Degrees of Freedom:   15
Residual Deviance: 149.4
```

Since Table 11.5 has 6 rows and 4 columns there are, as stated in the output, $5 \times 3 = 15$ degrees of freedom (d.f.). The value of G^2 is 149.4 which, when compared with a chi-squared distribution with 15 d.f., reveals that the table displays (as Figure 11.1 suggested) highly significant departures from independence.

The expected frequencies given here are also easily calculated from the marginal totals using Equation (4.2). For example, for murders in Texas using knives:

$$1133 \times 673/5279 = 144.4.$$

CHAPTER 12

LOG-LINEAR MODELS FOR $I \times J \times K$ TABLES

This chapter introduces the five types of model that are possible with three variables. It begins by extending the notation of Chapter 4:

- Denote the variables by A (having I categories), B (having J categories), and C (having K categories).
- Let p_{ijk} denote the probability of an observation belonging to cell (i, j, k) (i.e., to category i of variable A, category j of B, and category k of C).
- Let $v_{ijk} = \ln(Np_{ijk})$, where N is the total number of observations.
- Denote totals, as before, using zero subscripts. For example:

$$p_{ij0} = \sum_k p_{ijk}, \qquad p_{i00} = \sum_j \sum_k p_{ijk}, \qquad p_{000} = \sum_i \sum_j \sum_k p_{ijk}.$$

With three variables, A, B, and C, there are just nine models of interest. The relations between these models are illustrated in Figure 12.1. The restriction to these nine models is a consequence of the hierarchy constraint that is discussed in Chapter 13. There are five types of models corresponding to the five "levels" indicated in the figure. The model types are listed in Table 12.1.

At one extreme there is the mutual independence model, $A/B/C$, while at the other extreme there is the saturated model, which has as many parameters as there are cells in the contingency table (i.e., IJK) and therefore provides a

Categorical Data Analysis by Example, First Edition. Graham J. G. Upton.
© 2017 John Wiley & Sons, Inc. Published 2017 by John Wiley & Sons, Inc.

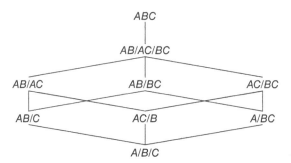

FIGURE 12.1 Model tree showing the connections between the nine models of possible interest in a three-way classification involving variables *A*, *B*, and *C*.

TABLE 12.1 Model types of possible interest with three variables, *A*, *B*, *C*

ABC	The interaction between any two variables is affected by the third variable. This is the *saturated model*.
AB/AC/BC	There are interactions between every pair of variables, but each interaction is unaffected by the category of the third variable.
AB/AC	There are interactions between *A* and *B* and between *A* and *C*. Neither interaction is affected by the category of the third variable. *Conditional independence between B and C, controlling for A.*
AB/C	There is an interaction between *A* and *B* that is unaffected by *C*. *C is independent of A and B.*
A/B/C	All pairs of variables independent of one another. The *mutual independence model*.

perfect fit to the data. The magnitudes of the parameter estimates in the saturated model often provide a good idea of which are the important interactions.

12.1 MUTUAL INDEPENDENCE: *A*/*B*/*C*

The mutual independence model will rarely apply, since the intention of collecting the data will have been to discover which variables are related and also the strength of those relations. However, the mutual independence model provides a useful basis for evaluating the benefits of more complex models.

The model states that each cell probability is the product of the corresponding (marginal) category probabilities:

$$p_{ijk} = p_{i00}p_{0j0}p_{00k}. \tag{12.1}$$

An example of a $2 \times 2 \times 2$ table displaying mutual independence is:

	C_1		C_2		
	B_1	B_2	B_1	B_2	Total
A_1	1	2	3	6	12
A_2	1	2	3	6	12
Total	2	4	6	12	24

In this example $p_{100} = \frac{1}{2}$, $p_{010} = \frac{1}{3}$, and $p_{001} = \frac{1}{4}$, so that, for example,

$$p_{111} = \frac{1}{2} \times \frac{1}{3} \times \frac{1}{4} = \frac{1}{24}.$$

With 24 observations in this fictitious example, the cell count for cell $(1,1,1)$ is $24 \times \frac{1}{24} = 1$, with similar calculations for the other cells.

As a log-linear model the mutual independence model would be written as

$$v_{ijk} = \ln(Np_{ijk}) = \mu + \lambda_i^A + \lambda_j^B + \lambda_k^C, \tag{12.2}$$

with appropriate constraints.

12.2 THE MODEL AB/C

An example of data for which this model would be a perfect fit is:

	C_1		C_2	
	B_1	B_2	B_1	B_2
A_1	1	2	3	6
A_2	3	4	9	12

Considering only observations belonging to category C_1, the odds on A_1 as opposed to A_2 are 1 to 3 for an individual belonging to category B_1, but 2 to 4 (equal to 1 to 2) for an individual belonging to category B_2. A convenient yardstick is the odds-ratio (see Section 3.5.3) which is equal to 1 for independence, but here is equal to

$$\frac{1/3}{2/4} = \frac{2}{3},$$

indicating that A and B are not independent.

Turning to category C_2, the odds on A_1 as opposed to A_2 are 3 to 9 for an individual belonging to category B_1, and 6 to 12 for an individual belonging to category B_2. The odds-ratio is therefore

$$\frac{3/9}{6/12} = \frac{2}{3}.$$

The odds-ratio associated with the categories of A and B is therefore the same for every category of C. This implies that the two-variable AB interaction is unaffected by C (in other words there is no ABC interaction).

To examine whether there is a relation between A and C alone, we rearrange the cells of the table:

	B_1		B_2	
	C_1	C_2	C_1	C_2
A_1	1	3	2	6
A_2	3	9	4	12

This time the odds-ratio in each subtable is equal to 1, implying that A and C are *conditionally independent* given B. The same holds true for B and C, so, for this table, a perfect fit is provided by the log-linear model

$$v_{ijk} = \ln(Np_{ijk}) = \mu + \lambda_i^A + \lambda_j^B + \lambda_k^C + \lambda_{ij}^{AB}, \qquad (12.3)$$

with appropriate constraints.

The lack of dependence of the category of C on those for A and B (and vice versa) is emphasized by rewriting the table as follows:

	A_1B_1	A_1B_2	A_2B_1	A_2B_2	Total
C_1	1	2	3	4	10
C_2	3	6	9	12	30
Total	4	8	12	16	40

In effect, this is a 2×4 table that displays independence between the variable C and the *compound variable* (AB). The marginal totals are included so that it is easy to see that, for example, the first entry 1 is equal to $4 \times 10/40$.

With this presentation, it is apparent that the cell probabilities are given by:

$$p_{ijk} = p_{ij0}p_{00k}. \qquad (12.4)$$

12.3 CONDITIONAL INDEPENDENCE AND INDEPENDENCE

In the previous example (repeated below for convenience), the value of θ (the odds-ratio) was the same for both subtables and the combined table:

	B_1		B_2					C_1	C_2
	C_1	C_2	C_1	C_2					
A_1	1	3	2	6	Ignoring B		A_1	3	9
A_2	3	9	4	12	$\Longrightarrow$		A_2	7	21
	$\theta = 1$		$\theta = 1$					$\theta = 1$	

Since $\theta = 1$ corresponds to independence, we can assert that each subtable displayed conditional independence between A and C while the combined table displayed independence between A and C. However, it is important to realize that conditional independence between two variables in subtables does not imply independence in the corresponding combined table. Here is an example:

	B_1		B_2					C_1	C_2
	C_1	C_2	C_1	C_2					
A_1	40	10	1	4	Ignoring B		A_1	41	14
A_2	20	5	20	80	$\Longrightarrow$		A_2	40	85
	$\theta = 1$		$\theta = 1$					$\theta \approx 6.2$	

The underlying reason for the huge difference between the odds-ratios for the subtable and that for the combined table is because of the impact of variable B on the other variables: there are very large AB and BC interactions. The importance of including all potentially influential variables in an analysis is a topic that we revisit in Section 12.6.

It is also the case that apparent independence in a combined table can be misleading when subtables are considered. Here is an example:

	C_1	C_2			B_1		B_2	
					C_1	C_2	C_1	C_2
A_1	3	9	Subdividing B		1	8	2	1
A_2	7	21	$\Longrightarrow$		6	1	1	20
	$\theta = 1$				$\theta = 1/48$		$\theta = 40$	

In this case, when B is ignored, it appears that A and C are independent. The reality is that there are strong AC interactions that vary enormously according to the category of B. This dependence of AC on B implies that there is a large ABC interaction (i.e., $\lambda_{ijk}^{ABC} \neq 0$).

12.4 THE MODEL AB/AC

An example of data for which this model would be a perfect fit is:

	C_1		C_2	
	B_1	B_2	B_1	B_2
A_1	1	2	3	6
A_2	2	5	4	10

Here the odds-ratios for the C_1 and C_2 subtables are

$$\frac{1/2}{2/5} = \frac{5}{4} \text{ and } \frac{3/6}{4/10} = \frac{5}{4}.$$

Since the odds-ratios are equal to one another, there is no ABC interaction (the AB interaction is not affected by C). Since the common value of the odds-ratios is not 1, A and B are not independent.

Since there is no ABC interaction, the odds-ratio between B and C will be the same for both subtables of A, and the odds-ratio between A and C will be the same for both subtables of B. These odds-ratios are, respectively, 1 and 2/3, indicating that while B and C are conditionally independent given A, A and C are not independent.

Because there is no BC interaction, the compound variables AB and AC are independent. We can see this by representing the data using two subtables:

	A_1B_1	A_1B_2	Total
A_1C_1	1	2	3
A_1C_2	3	6	9
Total	4	8	12

	A_2B_1	A_2B_2	Total
A_2C_1	2	5	7
A_2C_2	4	10	14
Total	6	15	21

In effect, the original 2^3 table separates into two disconnected 2^2 tables for the two categories of A. Within each subtable B and C are independent. For this model the cell probabilities are given by:

$$p_{ijk} = p_{ij0}p_{i0k}/p_{i00}. \tag{12.5}$$

As a log-linear model the model is

$$v_{ijk} = \ln(Np_{ijk}) = \mu + \lambda_i^A + \lambda_j^B + \lambda_k^C + \lambda_{ij}^{AB} + \lambda_{ik}^{AC}, \qquad (12.6)$$

with appropriate constraints.

12.5 THE MODELS *AB/AC/BC* AND *ABC*

These models do not have simple expressions for p_{ijk} in terms of totals of other cell probabilities. In log-linear terminology they are simple extensions of the *AB/AC* model. The *AB/AC/BC* model is

$$v_{ijk} = \ln(Np_{ijk}) = \mu + \lambda_i^A + \lambda_j^B + \lambda_k^C + \lambda_{ij}^{AB} + \lambda_{ik}^{AC} + \lambda_{jk}^{BC}, \qquad (12.7)$$

and the *ABC* model is

$$v_{ijk} = \ln(Np_{ijk}) = \mu + \lambda_i^A + \lambda_j^B + \lambda_k^C + \lambda_{ij}^{AB} + \lambda_{ik}^{AC} + \lambda_{jk}^{BC} + \lambda_{ijk}^{ABC}. \qquad (12.8)$$

Both models are subject to the usual constraints.

The model *AB/AC/BC* states that there are pairwise dependencies between the variables, with each dependency being unaffected by the third variable. Notice that if the interaction between *A* and *B* had been affected by *C*, then that would imply that there was a non-zero λ_{ijk}^{ABC} term in the model. In turn, if there is a non-zero λ_{ijk}^{ABC} term then that implies that the interaction between *A* and *C* is affected by *B* and the interaction between *B* and *C* is affected by *A*.

The model *ABC* is dominated by the three-variable interaction, λ_{ijk}^{ABC} since a non-zero λ_{ijk}^{ABC} implies that the interaction between any two of the variables is affected by the third variable.

12.6 SIMPSON'S PARADOX

In Section 12.3, we showed that when two subtables displaying conditional independence were combined, then the combined table need not display independence. We also showed that a combined table showing apparent independence, could result from subtables that showed marked departures from independence. Both situations have their counterparts in linear regression, as Figure 12.2 demonstrates. In the diagrams, independence is represented by a line parallel to the *x*-axis, where *y* is constant (so is unaffected by the value of *x*).

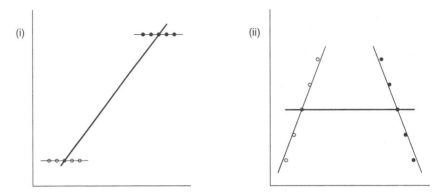

FIGURE 12.2 Graphical analogies of problems caused by ignoring a third variable. (a) Apparent dependence with conditional independence. (b) Apparent independence with strong dependence in the sub-populations.

When an influential third variable is mistakenly omitted, the results can be very misleading, as the following (fortunately hypothetical!) example demonstrates. In this scenario a doctor has to make the decision as to whether a critically ill patient should be given an injection or a pill, with the outcome being either recovery or death. Here are the historical figures for a patient with this type of illness:

	Recovery	Death
Injection	15	10
Pill	10	15

$$\theta = 2.25$$

Historically there have been 50 patients with 25 having received each treatment. The recovery rate is better for injections ($15/25 = 60\%$ than for the pill ($10/25 = 40\%$), so, in the absence of other information, the decision would be to inject the next unfortunate patient.

Fortunately, in this case the third variable is difficult to overlook since it is the sex of the patient. Here are the separate results for males and females:

	Males		Females	
	Recovery	Death	Recovery	Death
Injection	14	5	1	5
Pill	5	1	5	14
	$\theta = 0.56$		$\theta = 0.56$	

We now see that survival rates vary greatly between the sexes: $(14 + 5)/25 = 76\%$ for males, but only $(1 + 5)/25 = 24\%$ for females. This means that there is a very influential sex-outcome interaction.

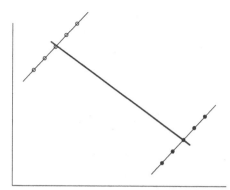

FIGURE 12.3   A graphical analogue of Simpson's paradox: positive slopes in sub-populations become a negative slope when the data are (incorrectly) treated as a single population.

When we look at the treatment results for males, we find that the survival rate for those given the pill ($5/6 \approx 83\%$) is greater than for those given the injection ($14/19 \approx 74\%$) implying that, for males, a pill is the preferred treatment.

When we look at the treatment results for females, we find that the survival rate for those given the pill ($5/19 \approx 26\%$) is greater than for those given the injection ($1/6 \approx 17\%$) implying that, for females, a pill is again the preferred treatment.

Thus, if we were to ignore the sex of the patient, we would choose to inject the patient, whereas taking account of the sex of the patient, we would use the pill, *regardless of the actual sex of the patient.*

This extraordinary result is termed Simpson's paradox as a result of an influential paper (Simpson, 1951). The graphical analogue is shown in Figure 12.3. Simpson's paradox is a form of *ecological fallacy.* The fallacy results from assuming that the conclusions from one level of aggregation apply to all levels of aggregation.

12.7 CONNECTION BETWEEN LOG-LINEAR MODELS AND LOGISTIC REGRESSION

Situations in which there is a single response variable have been the subject of the logistic models of earlier chapters. However, they could alternatively be handled using log-linear models. These models would be equally effective, and they would lead to the same goodness-of-fit statistics, parameter estimates, and fitted values (but they would be more cumbersome).

As an example, suppose that A and B are two explanatory variables, with C being the sole response variable. Suppose that A has I categories, B has J categories, and C has K categories. Now consider the three-variable saturated model given by Equation (12.8) and repeated here with p_{ijk} replaced by q_{ijk} to avoid notational confusion:

$$\ln(Nq_{ijk}) = \mu + \lambda_i^A + \lambda_j^B + \lambda_k^C + \lambda_{ij}^{AB} + \lambda_{ik}^{AC} + \lambda_{jk}^{BC} + \lambda_{ijk}^{ABC}, \quad (12.9)$$

with each set of parameters being subject to cornered or centered constraints.

Now suppose that $K = 2$ so that

$$\ln(Nq_{ij1}) = \mu + \lambda_i^A + \lambda_j^B + \lambda_1^C + \lambda_{ij}^{AB} + \lambda_{i1}^{AC} + \lambda_{j1}^{BC} + \lambda_{ij1}^{ABC},$$
$$\text{and } \ln(Nq_{ij2}) = \mu + \lambda_i^A + \lambda_j^B + \lambda_2^C + \lambda_{ij}^{AB} + \lambda_{i2}^{AC} + \lambda_{j2}^{BC} + \lambda_{ij2}^{ABC}.$$

Subtraction of one case from the other gives

$$\ln(q_{ij1}/q_{ij2}) = \{\lambda_1^C - \lambda_2^C\} + \{\lambda_{i1}^{AC} - \lambda_{i2}^{AC}\} + \{\lambda_{j1}^{BC} - \lambda_{j2}^{BC}\} + \{\lambda_{ij1}^{ABC} - \lambda_{ij2}^{ABC}\},$$

since terms not involving C cancel out.

The left-hand side is a log-odds, since, with $K = 2$, $q_{ij2} = 1 - q_{ij1}$. On the right-hand side there is a term not involving i or j, a term involving only i, a term involving only j, and a term involving both i and j. Such a model was given earlier by Equation (8.4) and is repeated here for convenience:

$$\ln(p_{ij}/(1 - p_{ij})) = \mu + \alpha_i + \beta_j + \theta_{ij}, \quad i = 2, \ldots, I, \quad j = 2, \ldots, J,$$

In this model μ is a measure of the typical cell frequency, α_i and β_j measure the effects of the specific categories i and j, and θ_{ij} is the effect on the response variable of the (i, j) category combination.

In a similar fashion it can be shown that every log-linear model involving a single response variable can be expressed as a logistic regression model. Of course, the reverse is also true.

Example 12.1   The dependence of political allegiance on gender and social class

In Example 10.2, we used logistic regression models to examine the relation between a single response variable (party preference in the United Kingdom, C) and two background factors: gender, A and social class, B. The data previously given as percentages in Table 10.4 are now reported as counts in Table 12.2.

TABLE 12.2 **Numbers supporting the three principal political parties, with subdivisions by gender and social class**

	Social classes A, B, and C1		Social classes C2, D, and E	
	Male	Female	Male	Female
Labour	188	196	183	128
Conservative	212	223	111	104
Liberal Democrat	46	52	39	18

Source: SN 6322, Gender and the Vote in Britain, 2007. Reproduced with permission of the UK Data Service.

We now use log-linear models (all containing the all-factor gender × class interaction) to re-examine the data. The results are shown in Figure 12.4, which should be compared with the earlier Figure 10.4.

The values of G^2 are very different, because now we are modeling 12 summary counts instead of the 1500 individual values. The AIC values here are therefore also quite different to the previous $AICc$ values. However, the actual values of either G^2 or AIC are not relevant. What matters is the differences in

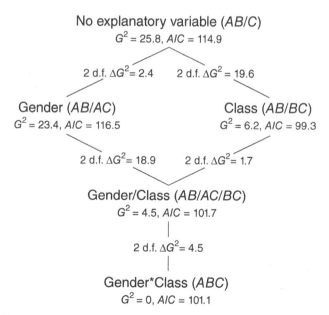

FIGURE 12.4 The fit of log-linear models describing the dependence of political allegiance on gender and/or class.

the G^2 values. These are identical to those obtained previously. In the same way, AIC again identifies the best fitting model as being (in the current notation) AB/BC.

REFERENCE

Simpson, E. H. (1951) The interpretation of interaction in contingency tables. *J. R. Stat. Soc. B*, **13**, 238–241.

CHAPTER 13

IMPLICATIONS AND USES OF BIRCH'S RESULT

Log-linear models were popularized in the early 1970s in a series of papers by Goodman (1970, 1971a, 1971b). This was before the introduction of the unifying concept of a single algorithm (Nelder and Wedderburn, 1972) to handle all linear models involving data from members of the exponential family of distributions (see Chapter 5). Prior to the use of the general linear model, an iterative procedure was used to fit log-linear models. The procedure, which had been introduced in a somewhat different context by Deming and Stephan (1940), has been rediscovered by several subsequent workers. The procedure is closely related to the result presented in Section 13.1.

13.1 BIRCH'S RESULT

Birch (1963) demonstrated that there was a connection between observed marginal totals, the maximum likelihood estimates of model parameters, and the marginal totals of the set of fitted frequencies based on those estimates. Expressed in our current notation the result is effectively this:

For each λ-parameter included in a log-linear model, the corresponding observed marginal total is exactly equaled by the corresponding marginal total of the fitted frequencies.

Categorical Data Analysis by Example, First Edition. Graham J. G. Upton.
© 2017 John Wiley & Sons, Inc. Published 2017 by John Wiley & Sons, Inc.

The iterative scaling procedure guarantees that this required equality of marginal totals is achieved.

13.2 ITERATIVE SCALING

Birch's result requires that relevant marginal totals should match. Iterative scaling provides a means of achieving these matches. The simple steps are as follows:

1. Begin with a working table of the same size (e.g., $I \times J \times K$) as that observed, but with every cell frequency equal to 1.
2. For a λ-parameter in the model, determine the corresponding marginal totals in both the observed table and the working table.
3. Scale the working table so that its marginal totals match those observed.
4. Repeat, cycling through the relevant λ-parameters until all totals match either exactly, or to a desired precision.

Fienberg (1970) proved that the procedure is sure to converge. In practice, convergence rarely takes many iterations. In cases where there is a simple expression that connects the fitted frequencies to appropriate totals of the observed frequencies (such as for the independence model, where $e_{ij} = f_{i0}f_{0j}/f_{00}$) convergence occurs after just one iteration. These models are called *direct models*.

For models that are not direct models, the convergence is not only assured, but fast. Since the method requires only multiplication, with no awkward matrix manipulations, using iterative scaling was an attractive method for fitting log-linear models when computer power was limited.

A rather curious consequence of the procedure is that the fitted frequencies are obtained without having calculated parameter estimates. If parameter estimates are required then the general linear model should be used. If using R, this implies using the `glm` command, rather than the `loglm` command.

Example 13.1 Suppose that we wish to fit the independence model to the following 2×2 table:

	B_1	B_2	Row marginal total: f_{i0}
A_1	$f_{11} = 10$	$f_{12} = 20$	$f_{10} = 30$
A_2	$f_{21} = 30$	$f_{22} = 40$	$f_{20} = 70$

Column marginal total: f_{0j} $f_{01} = 40$ $f_{02} = 60$ Grand total: $f_{00} = 100$

Since the independence model is

$$v_{ij} = \mu + \lambda_i^A + \lambda_j^B, \qquad i = 1, 2; j = 1, 2,$$

the λ-parameters of interest are the $\{\lambda_i^A\}$ and the $\{\lambda_j^B\}$ and the corresponding marginal totals are respectively the $\{f_{i0}\}$ and the $\{f_{0j}\}$. Denoting fitted frequencies by $\{e_{ij}\}$, with the usual notation for totals, the independence model implies that

$$e_{i0} = f_{i0} \text{ for all } i \text{ (because } \lambda_i^A \text{ is in the model),}$$

and $e_{0j} = f_{0j}$ for all j (because λ_j^B is in the model).

It does not matter in what order totals are matched; the same end results will be obtained. We will arbitrarily start with the $\{\lambda_i^A\}$, so that we are interested in the row marginal totals.

$B_1 \; B_2$ Marginal total

	B_1	B_2	Marginal total	
A_1	1	1	2	needs to be 30, so $\times$ by 15 $\rightarrow$ 15 15
A_2	1	1	2	needs to be 70, so $\times$ by 35 $\rightarrow$ 35 35

Next we consider the $\{\lambda_j^B\}$:

	B_1	B_2
A_1	15	15
A_2	35	35
Marginal total:	50	50
Should be:	40	60
Therefore $\times$ by:	40/50	60/50
	$\downarrow$	$\downarrow$
	12	18
	28	42

In this case one iteration has been sufficient, since both sets of scaled margins match those observed. The independence model is the simplest example of a direct model.

13.3 THE HIERARCHY CONSTRAINT

Consider a cross-classification involving three variables, A, B, and C. Suppose that we find that the independence model $A/B/C$ does not fit the data well, whereas the model AB/C is an acceptable fit. In the previous notation, the model AB/C is

$$v_{ijk} = \ln(Np_{ijk}) = \mu + \lambda_i^A + \lambda_j^B + \lambda_k^C + \lambda_{ij}^{AB}.$$

We might feel that it would be worth testing the fit to the simpler model

$$v_{ijk} = \ln(Np_{ijk}) = \mu + \lambda_k^C + \lambda_{ij}^{AB},$$

even though we previously argued that if there was an important AB interaction, then this implied that the categories of A and B mattered. We now demonstrate that Birch's result implies that, if a model contains λ_{ij}^{AB}, then the model "must" include λ_i^A and λ_j^B. This is easy to demonstrate as follows:

$$\lambda_{ij} \text{ in the model} \implies e_{ij0} = f_{ij0}$$

$$\text{Summing over } j \implies e_{i00} = f_{i00}$$

$$\text{But } e_{i00} = f_{i00} \implies \lambda_i^A \text{ is in the model.}$$

The general result (the hierarchy constraint) is:

If a multi-variable interaction I appears in a model, then all the interactions involving subsets of I must also appear in the model.

Thus, for example, if λ_{ijk}^{ABC} appears in the model, then so must λ_{ij}^{AB}, λ_{ik}^{AC}, and λ_{jk}^{BC} (and λ_i^A, λ_j^B, λ_k^C, and μ). Thus log-linear models are completely specified by listing their most complex interactions or effects.

13.4 INCLUSION OF THE ALL-FACTOR INTERACTION

Consider a general situation involving the cross-tabulation of m response variables $(R_1, R_2, \ldots, R_m)$ and n background factor variables $(F_1, F_2, \ldots, F_n)$. Suppose we were to attempt to fit the model of independence across all $(m + n)$ variables. There would be three possible contributions to any lack of fit:

The response variables are not independent of one another	Not uninteresting
The response variables are dependent on the background variables	Of prime interest
The background factors are not independent of one another	Irrelevant

Suppose we interview a sample of people concerning their preferences from a range of breakfast cereals. The fact that 60% of the females were over 50, whereas only 47% of males fell in the same age bracket, provides no information concerning breakfast cereals. It is a fact of the sampling process and should in no way contribute to any lack of fit for whatever model is being proposed.

We want to ensure that there is an exact match between the observed and fitted numbers of females aged over 50 and likewise for the other age-sex combinations. The match is achieved by using Birch's result: we simply insist that the age-sex interaction is included in every model considered. This ensures that the relevant observed marginal totals are exactly reproduced and any lack of fit involves the response variables.

Returning to the general case, the simplest log-linear model of interest will therefore be the model

$$R_1/R_2/\cdots/R_m/F_1F_2\cdots F_n,$$

which states that the response variables are all mutually independent of one another and of the background factors. Every model subsequently considered will include the all-factor interaction $F_1F_2\cdots F_n$.

13.5 MOSTELLERIZING

This term was introduced by Upton (1978) as an alternative to "standardizing"which was the term used by Mosteller (1968) in his careful description of the use of iterative scaling in the context of the adjustment of a sample or survey to match known population figures.

The process is equally effective with tables of any dimension. It has been used in market research (Upton, 1987), where it is known as *raking* or *rimweighting*. In the context of traffic flow it is variously known as the *Cross-Fratar procedure* (Fratar, 1954) or the *Furness method* (Furness, 1965).

Underlying this application of the iterative scaling algorithm is the assumption that any biases are limited to misrepresentation of the frequencies of the categories of the classifying variables, with category combinations being otherwise fairly represented. In terms of model parameters, the assumptions are that the "main effect"λ-parameters for single variables may be incorrect, but those for interactions are correctly measured. This means that the interaction parameters for the saturated model applied to the original data will have identical values to those of the interaction parameters for the saturated model applied to the Mostellerized data.

TABLE 13.1 Political preferences and genders of respondents as recorded by two researchers

	Experienced researcher				Novice		
	Party A	Party B	Total		Party A	Party B	Total
Male	20	90	110	Male	30	70	100
Female	60	30	90	Female	220	80	300
Total	80	120	200	Total	250	150	400

Example 13.2 In this fictitious example, we suppose that two researchers are asked to sample the population to find out whether the political opinions of males differ from those of females. The researchers are told to choose random members of the population and to ask the question "If you had to choose Party A or Party B to form the next government, which would you choose?". One of the researchers is an experienced member of the survey team, but the other is a novice, a young man for whom this is his first task. Table 13.1 records the results of the two researchers.

It is immediately apparent that the novice has not been sampling at random! If we are hoping to deduce the proportion of the population that consists of female supporters of Party A, then, on their own, the novice's results are useless (and those of the experienced researcher are a little suspect). However, if we know that 55% of the population prefer Party A, and we know that 51% of the population are female, then iterative scaling can recover the desired

TABLE 13.2 The first two complete iterations to recover information from the novice's sample

					First iteration						
	Raw data				Row scaled				Column scaled		
	A	B	Total		A	B	Total		A	B	Total
Male	30	70	100		14.7	34.3	49		15.5	32.2	47.7
Female	220	80	300	→	37.4	13.6	51	→	39.5	12.8	52.3
Total	250	150	400		52.1	47.9	100		55	45	100

					Second iteration						
					Row scaled				Column scaled		
					A	B	Total		A	B	Total
					15.9	33.1	49		16.1	32.7	48.8
				→	38.5	12.5	51	→	38.9	12.3	51.2
					54.4	45.6	100		55	45	100

estimates. Table 13.2 shows the first two complete iterations when scaling the novice's data. The speed of convergence is apparent. It appears that about 39% of the population are female supporters of Party A.

REFERENCES

Birch, M. W. (1963) Maximum likelihood in three-way contingency tables. *J. Roy. Statist. Soc., B*, **25**, 220–233.

Deming, W. E. and Stephan, F. F. (1940) On a least squares adjustment of a sampled frequency table when the expected marginal totals are known. *Ann. Math. Statist.*, **11**, 427–444.

Fienberg, S. E. (1970) An iterative procedure for estimation in contingency tables. *Ann. Math. Stat.*, **41**, 907–917.

Fratar, T. (1954) Vehicular trip generation by successive approximations. *Traffic Quarterly*, **8**, 53–65.

Furness, K. P. (1965) Time function iteration. *Traffic Engineering Control*, **7**, 458–460.

Goodman, L. A. (1970) The multivariate analysis of qualitative data: interactions among multiple classifications. *J. Amer. Stat. Ass.*, **65**, 226–256.

Goodman, L. A. (1971a) The analysis of multidimensional contingency tables, stepwise procedures and direct estimation methods for building models for multiple classifications. *Technometrics*, **13**, 33–61.

Goodman, L. A. (1971b) Partitioning of chi-square, analysis of marginal contingency tables and estimation of expected frequencies in multidimensional contingency tables. *J. Amer. Statist. Assoc.*, **66**, 339–344.

Mosteller, C. F. (1968) Association and estimation in contingency tabes. *J. Amer. Statist. Assoc.*, **63**, 1–28.

Nelder, J. A., and Wedderburn, R. W. M. (1972) Generalized linear models. *J. R. Stat. Soc. A*, **135**, 370–384.

Upton, G. J. G. (1978) *The Analysis of Cross-tabulated Data*, John Wiley & Sons, Chichester.

Upton, G. J. G. (1987) On the use of rim weighting. *J. Market Res. Soc.*, **29**, 363–366.

CHAPTER 14

MODEL SELECTION FOR LOG-LINEAR MODELS

The techniques outlined in Chapter 9 continue to apply, with stepwise methods providing an invaluable aid when there are many variables.

The amount of data available is again relevant as the examples will demonstrate. If there is a huge amount of data then any unsaturated model may have a "significantly" poor fit. On the other hand, when data are scarce, a cross-classification involving several variables may lead to a large number of zero cell frequencies and a difficulty in finding models that do not fit the data acceptably!

With any data analysis it is useful to begin by "looking at the data." In the context of log-linear models, this suggests examining the magnitudes of the possible interactions using a cobweb diagram to provide a pictorial guide to the magnitudes of two-variable interactions and studying the magnitudes of the parameters in the saturated model to gain an idea of which other interactions may be relevant.

14.1 THREE VARIABLES

When there are just three variables, model selection is straightforward, since there are at most nine models to consider (see Figure 12.1). When two of the variables are background factors, the number reduces to five, as illustrated in Figure 12.4.

Categorical Data Analysis by Example, First Edition. Graham J. G. Upton.
© 2017 John Wiley & Sons, Inc. Published 2017 by John Wiley & Sons, Inc.

Example 14.1 Belief in the existence of God and an afterlife

Table 11.1 presented data from a cohort of individuals born in the United Kingdom during a single week in 1970. Re-interviewed in 2012, the proportion of females who claimed to believe in God was found to be significantly greater than the proportion of males who believed in God. We now re-examine the data using four categories rather than two, and including information on the respondents' belief in an afterlife. These more detailed data are given in Table 14.1.

There are some very strong interactions present. While these may not be immediately apparent from the table of numbers (too many numbers!), they are certainly apparent in the cobweb diagram shown in Figure 14.1. The strongest positive associations link belief in God with belief in an afterlife (and the corresponding disbeliefs).

To get a full picture we now fit some log-linear models, starting with an examination of the parameter estimates for the saturated model:

R code

```
df<-as.data.frame(table(God,After,Sex));
summary(glm(Freq~Sex*God*After,data=df,family=poisson))
```

An extract from the results is given in Table 14.2 which shows the most significant differences within each group of interaction parameters. For simplicity of exposition we will denote the variables by A for belief in an afterlife, B for belief in God, and C for sex. There appears to be no AC interaction, but

TABLE 14.1 Replies given by members of the 1970 British Cohort Study when interviewed in 2012 concerning their belief in God and in afterlife

	Males				Females			
	Yes	Yes?	No?	No	Yes	Yes?	No?	No
There is a God	233	72	18	23	471	144	42	24
There may be a God	112	420	365	56	328	824	447	46
Do not know if there is a God	25	154	622	165	53	229	479	72
There is not a God	27	70	342	747	66	123	254	248

Source: SN 7473, 1970 British Cohort Study: Forty-Two-Year Follow-Up, 2012. Reproduced with permission of the UK Data Service.

In the table the question marks indicate uncertainty: thus Yes? implies that the respondent thought there might be an afterlife, while No? indicates that the respondent thought that there might not be an afterlife. By contrast, those answering Yes or No were certain.

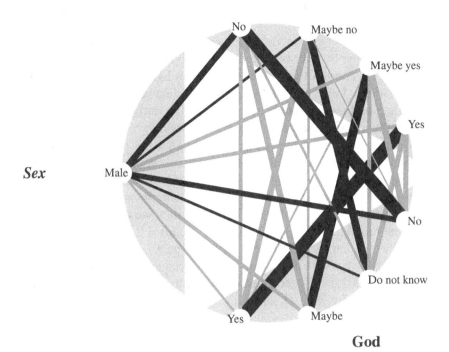

Afterlife

Sex

God

FIGURE 14.1 A cobweb diagram showing the relative importance of the various two-variable interactions for the data of Table 14.1.

extremely strong AB and BC interactions, suggesting that the model AB/BC could be appropriate. For this model, which has 12 degrees of freedom, the value of G^2 is 87.2, which suggests a very significant lack of fit. However, the next best model, AB/AC, which has the same number of degrees of freedom,

TABLE 14.2 The most significant parameter estimates (using cornered constraints) for each group of interaction parameters in the saturated model fitted to the data of Table 14.1. The variables are Belief in God (A), Belief in the afterlife (B), and Sex (C)

Interaction	Est.	S. E.	z	Tail Prob.	Parameter
SexFemale:GodYes	-0.19	0.24	-0.8	0.43	AC
SexFemale:AfterNo	-2.00	0.24	-8.3	†	BC
GodYes:AfterNo	-5.63	0.29	-19.2	†	AB
SexFemale:GodYes:AfterNo	1.34	0.39	3.5	0.0005	ABC

†, probability less than 2×10^{-16}.

has $G^2 = 318.1$. The model that states that a person's sex has no bearing on their belief in either God or the afterlife (model AB/C) has an extra 3 degrees of freedom, but $G^2 = 736.1$.

So what is the "best" model? The saturated model is of course the only model that perfectly describes the data. However, we can reasonably assume that a different cohort of individuals would not have resulted in precisely the figures given in Table 14.1. Indeed, no set of data can be expected to match *any* of the unsaturated models in this book. The entire modeling exercise is no more than a procedure for understanding which relationships are strong and which are weak, or barely exist. Although a tail probability of 0.0005 (the value shown in Table 14.2) is very small, it is very much greater than the probabilities associated with the AB and BC interactions. These tiny probabilities are partly a consequence of the large sample size. If we toss a coin for which P(Head) = 0.5001 a million times, then we can expect to obtain a significant rebuttal of the hypothesis that P(Head) = 0.5, but will we care? To see whether we care in the present context, Table 14.3 gives the last row of Table 14.1 together with the (rounded) fitted frequencies for competing models. Agresti (2007) suggested that a useful informal measure of fit is provided by the *dissimilarity index D*, where

$$D = 50 \times \sum |e - f| / \sum f, \qquad (14.1)$$

and where f denotes an observed frequency, e denotes the corresponding value given by a model, and the summation is over all cells in the table. The value of D can be interpreted as the percentage of the observations that would need to move to a different cell of the table in order to achieve a perfect fit. The values of D are given as the final column of the table.

We might feel that the AB/BC model gives an acceptable fit, whereas AB/C clearly does not. Thus the BC interaction cannot be ignored: the implication

TABLE 14.3 Observed and fitted frequencies concerning belief in the afterlife for respondents who claimed not to believe in God, together with the values of D for the fits to the entire table

	Males				Females				D
	Yes	Yes?	No?	No	Yes	Yes?	No?	No	
Observed	27	70	342	747	66	123	254	248	
ABC	27	70	342	747	66	123	254	248	0
$AB/AC/BC$	35	83	341	727	58	110	255	268	2.1
AB/BC	28	68	312	714	65	125	284	281	4.1
AB/C	44	91	282	470	49	102	314	525	13.7

is that males differ very significantly from females in the extent to which they are likely to believe in God.

14.2 MORE THAN THREE VARIABLES

As the number of variables increases, so the number of possible models rises rapidly. Precisely how many models are possible in a given situation will depend on the separate numbers of response variables and background factors. The number is restricted by the need to include the all-factor interaction in every model. This restriction occurs automatically with logistic models and ensures that any lack of fit represents the model's inability to describe the relations between the response variables and the factors (and between the response variables themselves).

Example 14.2   The hands of blues guitarists

This example is concerned with the possibility that early twentieth-century African-American blues guitarists developed distinctive regional styles as a result of the way in which they plucked the strings. A fascinating study by Andrew Cohen, published in the Winter 1996 edition of *American Music*, identified four variables of interest:

A	Date of birth	Before 1906 (primarily "songsters"), or After 1905 (primarily "bluesmen")
B	Region	West (mainly TX), Central (mainly MS), and East (mainly GA)
C	Thumb style	Cohen used three classes: Alternate (between thumb and finger), Utility (thumb used only when required), and Dead
D	Hand posture	The relative positions of thumb and forefinger. Cohen used three classes: Extended, Stacked, and Lutiform

Cohen collated information on 93 guitarists. A cross-tabulation for the four variables above (with categories in the order shown) is given in Table 14.4.

With 54 cells and 93 observations it is no surprise that 21 of the cell frequencies are zero. Based on the birthdays it would appear that playing of the blues originated in the east and moved west in later years. However date and region are the factors in this data set, so the simplest model of interest is $AB/C/D$.

TABLE 14.4 The thumb styles and hand positions of 94 blues guitarists in the southern states of the United States

Before 1906	West			Central			East		
	Alt.	Util.	Dead	Alt.	Util.	Dead	Alt.	Util.	Dead
Extended	1	0	2	5	5	4	4	0	3
Stacked	0	0	2	2	2	5	1	1	1
Lutiform	0	0	1	0	1	2	0	0	0
After 1905									
Extended	14	4	2	2	4	0	0	1	0
Stacked	0	3	1	0	4	5	0	1	2
Lutiform	1	0	0	0	3	3	0	0	1

Source: Cohen, 1996. Reproduced with permission of University of Illinois Press.

We begin, as usual, with a cobweb diagram (Figure 14.2) which suggests that the strongest association concerns those using the alternate thumb style who are disproportionately likely to be using the extended hand position. This happens particularly in the western region and in the later years. This finding reflects the fact that by far the largest frequency corresponds to that particular combination.

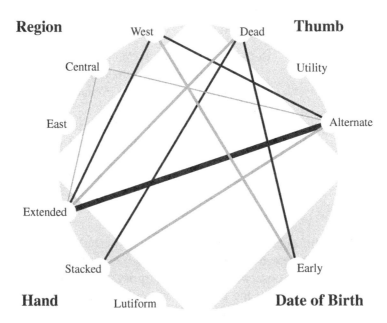

FIGURE 14.2 A cobweb diagram showing the relative importance of the various two-variable interactions for the data of Table 14.4.

On this occasion, because of the combination of small cell frequencies and large numbers of cells, the saturated model is not very informative. Instead we use the stepwise procedures discussed in Section 9.3.

R code

```
# Thumb, Region, Hand, DateOfBirth, Freq are vectors
df<-data.frame(expand.grid(Thumb=Thumb,Region=Region,
    Hand=Hand, DateOfBirth=DateOfBirth),Freq,
    stringsAsFactors=TRUE);
step(glm(Freq~Thumb*Region*Hand*DateOfBirth,
    data=df,family=poisson))
```

The automated stepwise procedure uses AIC as its criterion and terminates at the model ACD/ABD. However, the stepwise procedure only looks one step ahead in the model tree. In Section 9.3, we saw that it is sometimes possible to find a model with a smaller AIC value by considering models two steps (or more) through the model tree. The current data set provides another example where this is possible.

An extract from the model tree is shown in Figure 14.3. The automated procedure stopped at ABC/ACD because its AIC value (168.2) was less than that for the two possible simplifications $ABC/AD/CD$ (169.1) and $AB/BC/ACD$ (175.8). The difference in the G^2-values for ABC/ACD and $ABC/AD/CD$ is $27.47 - 18.59 = 8.88$. This is associated with $28 - 24 = 4$ degrees of freedom. The probability of a χ_4^2-variable exceeding 8.88 by chance is about 7%, suggesting that, despite its larger AIC value, $ABC/AD/CD$ should be given serious consideration as a model providing a simpler explanation of the data. Before considering that model, however, it makes sense to investigate whether any further simplification is feasible. As Figure 14.3 shows, further simplification is indeed possible, since removing the AD interaction (to give the model ABC/CD) releases 2 degrees of freedom, while G^2 increases by just 1.71 and, as a consequence, the AIC value falls to a new low (166.8).

The possible simplifications of the model ABC/CD result in unacceptably large increases in both G^2 and AIC, so this is the final model. To understand why a set of parameters are required it can be helpful to compare the expected frequencies resulting from the original model and from the rejected simplification. In this case, since the models ABC/CD and $AB/AC/BC/CD$ do not differ with respect to variable D, the differences can be conveniently summed over D; the results are shown in Table 14.5. The ABC interaction is evidently required to deal with the manner in which the combination of an alternate thumb style and an extended hand position varied with region across the time periods.

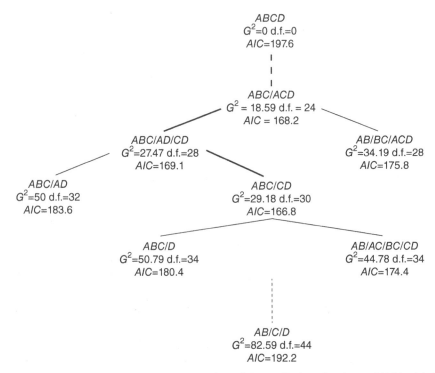

FIGURE 14.3 Extract from the tree of models applied to the data of Table 14.4. The variables are *A*, Date of birth; *B*, Region; *C*, Thumb style; *D*, Hand posture.

Finally, to understand the *CD* interaction we can compare the expected frequencies for the models *ABC/D* and *ABC/CD*, summing over variables *A* and *B*. The results are shown in Table 14.6. Once again it is the preference for an extended hand position with alternate thumb use that stands out from the data.

Comparison of the fitted frequencies for a pair of nested models is always useful when attempting to understand the relationships in a set of data.

TABLE 14.5 **The expected frequencies (summed over *D*) for the model *AB/AC/BC/CD* subtracted from those for the model *ABC/CD***

	West			Central			East		
	Alt.	Util.	Dead	Alt.	Util.	Dead	Alt.	Util.	Dead
Before 1906	−2.6	−0.6	3.1	1.0	0.8	−1.9	1.3	−0.2	−1.1
After 1905	2.6	0.6	−3.1	−1.0	−0.8	1.9	−1.3	0.2	1.1

TABLE 14.6   The expected frequencies (summed over D) for the model ABC/D subtracted from those for the model ABC/CD

	Alternate	Utility	Dead
Extended	9.5	−1.9	−7.6
Stacked	−6.7	1.6	5.0
Lutile	−2.9	0.3	2.6

REFERENCE

Agresti, A. (2007) *An Introduction to Categorical Data Analysis*, 2nd ed., John Wiley & Sons, Inc., Hoboken, NJ.

CHAPTER 15

INCOMPLETE TABLES, DUMMY VARIABLES, AND OUTLIERS

15.1 INCOMPLETE TABLES

In a cross-tabulation involving many cells, but with only a limited amount of data, there is a good chance that some cell frequencies will be zero. Such zeroes are called *random zeroes* because they are a consequence of the (hopefully) random sampling that occurred. Sometimes, however, there are so-called *structural zeroes* where a cell has a zero count by design. The resulting table is called an *incomplete table*.

Example 15.1   Health concerns of teenagers

Table 15.1 presents data on the health concerns of teenagers. The zeroes in this incomplete table can be anticipated, since male teenagers will not have menstrual problems.

15.1.1 Degrees of Freedom

When there are structural zeroes present, the number of degrees of freedom is usually the number that would have applied if there had been no structural zeroes, reduced by the number of structural zeroes.

Categorical Data Analysis by Example, First Edition. Graham J. G. Upton.
© 2017 John Wiley & Sons, Inc. Published 2017 by John Wiley & Sons, Inc.

TABLE 15.1   The health concerns of teenagers

		Personal health concerns			
Sex	Age	Sex, reproduction	Menstrual problems	How healthy I am	Nothing
Male	12–15	4	0	42	57
	16–17	2	0	7	20
Female	12–15	9	4	19	71
	16–17	7	8	10	31

Source: Brunswick, A. F. (1971) Adolescent health, sex, and fertility. *Amer. J. Pub. Health*, **61**, 711–720.

Example 15.1   Health concerns of teenagers (*continued*)

In this case, the obvious question is whether, leaving the implausible male menstrual concerns cells on one side, the health concerns (A) are otherwise independent of sex (B) and age (C). The simplest model of interest (since B and C are background factors) is A/BC, which, for a complete $2 \times 2 \times 4$ table would have 9 degrees of freedom. In this case, therefore, with two impossible cells, it has $9 - 2 = 7$ degrees of freedom.

The exception to the rule that the degrees of freedom are reduced by the number of structural zeroes occurs when there are so many structural zeroes, that a structurally zero marginal total is formed. In such a case, for each relevant structural zero marginal total, one degree of freedom is recovered. A marginal total is relevant if it would be exactly fitted as a consequence of Birch's result (Section 13.1).

Example 15.2   Eyesight quality in England

Table 15.2 presents a pathological example of a zero marginal total. Fitting the independence model to the 3×3 table as it stands, there will be

TABLE 15.2   Cross-tabulation of questions concerning respondents' eyesight

	Whether certified partially sighted or blind			
Self-reported eyesight	Yes, partially sighted	Yes, blind	No	Total
Good	0	0	0	0
Fair	17	0	592	609
Poor	20	3	133	156

Source: SN 7649, Health Survey for England, 2013. Reproduced with permission of the UK Data Service.

$(3 - 1)(3 - 1) - 3$ (for the structural zeroes) $+1$ (for the structural marginal total) $= 2$ degrees of freedom. Of course, any rational analysis would commence by discarding the first row, to leave a 2×3 table, thus confirming that there are $(2 - 1)(3 - 1) = 2$ degrees of freedom.

15.2 QUASI-INDEPENDENCE

In cases where there are no structural zeroes, the simplest model that might describe a data set is usually the independence model. When there are structural zeroes, it might be the case that independence would still describe the relationship between the variables, if the structural zeroes were ignored. This is the *quasi-independence model*.

Table 15.3 shows an artificial data set that provides a perfect fit to the two-variable quasi-independence model: in each complete column the cells are in the ratio 4 to 2 to 1. Notice that, because of the structural zero, it is no longer the case that the expected frequencies under independence are a simple function of the row, column, and overall totals.

TABLE 15.3 An artificial table displaying a perfect fit to the quasi-independence model. The top left cell is a structural zero

	B_1	B_2	B_3	B_4	B_5	Total
A_1	0	100	20	60	40	220
A_2	10	50	10	30	20	120
A_3	5	25	5	15	10	60
Total	15	175	35	105	70	400

15.3 DUMMY VARIABLES

Whenever there are subsets of the cells in a cross-tabulation that are of special interest, or that can be expected to behave in a different way from the remaining cells, it can be useful to invoke one or more dummy variables. We will find applications of dummy variables both in this chapter and the next. We begin with the trivial example of an incomplete table.

Example 15.1 Health concerns of teenagers (*continued*)

There were two individual cells in the cross-classification that we regarded as impossible. Nevertheless, it is not completely inconceivable that a very ignorant (or mischievous) young male might claim to have menstrual problems. We can address this possibility by introducing a dummy variable D

TABLE 15.4 The data of Table 15.1 presented as an incomplete $2^2 \times 4 \times 3$ table

D_1				D_2				D_3			
4	×	42	57	×	0	×	×	×	×	×	×
2	×	7	20	×	×	×	×	×	0	×	×
9	4	19	71	×	×	×	×	×	×	×	×
7	8	10	31	×	×	×	×	×	×	×	×
Total = 291				Total = 0				Total = 0			

having three categories: D_1, realistic cells; D_2, 12–15-year-old male claiming to worry about menstrual problems; D_3, 16–17-year-old male claiming to worry about menstrual problems.

Since categories D_2 and D_3 identify single cells, their marginal frequencies will be equal to the individual cell counts and, as a consequence of Birch's result (Section 13.1), their frequencies (in this case, both zeroes) will be exactly reproduced. The original $2^2 \times 4$ table becomes the highly incomplete $2^2 \times 4 \times 3$ table, given as Table 15.4.

Notice that the two degrees of freedom "lost" to the structural zeroes of the original table are now, in effect, allocated to the dummy variable. A more succinct representation of the distribution of the categories of the dummy variable is provided by Table 15.5.

TABLE 15.5 The categories of the dummy variable D for the data of Table 15.1

1	2	1	1
1	3	1	1
1	1	1	1
1	1	1	1

15.4 DETECTION OF OUTLIERS

In the context of cross-classified data, an outlier is a category combination that does not match the pattern suggested by other category combinations. An example is provided by Table 15.6 which is a table that displays perfect independence, with the exception of the top-left category combination, where the observed count of 2000 is 100 times greater than that suggested by the counts in the remainder of the table. The following model provides a perfect fit to the data:

$$v_{ij} = \mu + \lambda_i^A + \lambda_j^B + \delta_{ij}\lambda^D, \tag{15.1}$$

TABLE 15.6   An artificial table in which the top-left cell is an obvious outlier

	B_1	B_2	B_3	B_4	B_5	Total
A_1	2000	100	20	60	40	2220
A_2	10	50	10	30	20	120
A_3	5	25	5	15	10	60
Total	2015	175	35	105	70	2400

where

$$\delta_{11} = 1 \text{ and } \delta_{ij} = 0 \text{ otherwise.}$$

A significance test of whether cell (1,1) is an outlier is provided by testing whether λ^D differs significantly from zero. If $\lambda^D = 0$, then, in the present situation, the model simplifies to the independence model. Suppose that the value of the G^2 statistic for this model is G_0^2, while the value for the model including the dummy variable is G_1^2. Then the difference $G_0^2 - G_1^2$ is associated with the 1 degree of freedom that results from the addition of the single λ^D-parameter.

Now consider a general case, involving any model, and any number of variables. Let the model of interest be denoted by M_0, with goodness-of-fit G_0^2. Let $M_1(C)$ be the same model augmented by a dummy parameter that identifies the specific cell C as an outlier. Suppose this model has goodness-of-fit $G_1^2(C)$. As usual, the two goodness-of-fit values have asymptotic chi-square distributions, so the difference $G_0^2 - G_1^2(C)$ would be compared with a χ_1^2-distribution.

However, the comparison with a χ_1^2-distribution is only valid if cell C has been identified prior to studying the data. Usually we will not know in advance whether there is an outlier. With a cross-classification comprising N cells, a natural procedure would be to examine each cell in turn in case it might be an outlier. The prime candidate would then be the cell for which the difference $G_0^2 - G_1^2(C)$ was greatest. Since we are giving ourselves N chances of finding a large value, the value we find should be compared not with a χ_1^2-distribution but with the distribution of the largest of N observations from a χ_1^2-distribution.

Note that, if there is an outlier cell, then its removal may result in a much simpler explanation of the relationships between the remaining variables.

Example 15.3   Visiting habits of Aberdeen mothers-to-be

Table 15.7 reports information on the extent to which 87 child-bearing women from working-class families in Aberdeen visited friends and an antenatal

TABLE 15.7 Visits to the antenatal clinic, and to friends, by working-class mothers in Aberdeen

Antenatal clinic	Visits to friends	First child		Not first child	
		Walked	Used bus	Walked	Used bus
Regular use	Daily	5	13	14	2
	At least once a week	6	6	6	6
	Less than once a week	0	4	2	5
Little use	Daily	1	0	16	0
	At least once a week	6	2	13	10
	Less than once a week	2	0	3	5

Source: McKinlay, 1973. Reproduced with permission of Oxford University Press.
Note that some respondents gave information about visits to more than one set of friends.

clinic. There are two factors: whether or not the child will be a first child (*A*), and whether the friends are visited by walking or by bus (*B*). The two response variables are whether or not the mother regularly attends an antenatal clinic (*C*), and the regularity with which the mother visits her friends (*D*).

Figure 15.1 presents the cobweb diagram for these data. It appears that there are appreciable *AB*, *AC*, *BC*, *BD*, and *CD* interactions. Stepwise

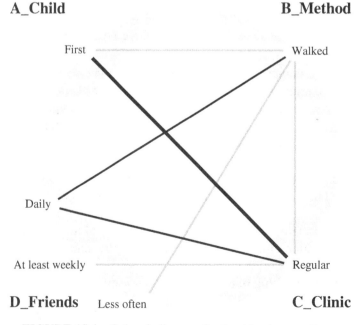

FIGURE 15.1 Cobweb diagram for the Aberdeen mothers.

methods (see Chapter 9) confirm this impression with the selected model being $AB/AC/BC/BD/CD$. This model has 11 degrees of freedom with $G^2 = 29.30$ corresponding to a tail probability of about 0.2%. The following R code stores the G^2-value in gsq0.

R code

```
C<-c("Regular","Little");
D<-c("Daily","Weekly","Less");
A<-c("First","Not first");
B<-c("Walk","Bus");
Freq<-c(5,13,14,2,6,6, etc,2,13,10,2,0,3,5);
df<-data.frame(expand.grid(B=B,A=A,D=D,C=C),
     Freq,stringsAsFactors = TRUE);
gsq0<-glm(Freq~A*B+A*C+B*C+B*D+C*D,
     data=df,family="poisson")$deviance;
```

A convenient method of assessing whether a cell is an outlier is to examine the change in G^2 that results from repeating the analysis with the chosen cell given zero weight. The following R code achieves this for each cell in turn, with the maximum change being stored in maxgdiff.

R code (*continued*)

```
n=length(Freq);
gdiff=0*c(1:n);
for(i in 1:n){
     wt<-c(rep(1,times=n));
     wt[i]<-0;
     gdiff[i]<-gsq0-glm(Freq~A*B+A*C+B*C+B*D+C*D,
          data=df,weights=wt,family="poisson")$deviance;
     }
maxgdiff<-max(gdiff)
maxgdiff;
```

The maximum change in G^2 is found to be 14.4, with the source being the count of 13 in the first row of the data. To assess whether this maximum change is unusually large, we use simulation. The following R code generates 99,999 samples of 24 observations from a χ_1^2-distribution. The code then determines the largest value in each set, counts the number of these largest values that are at least as great as the value observed, and presents the result as a tail probability. Because we are using simulations, we will get different results with each run of the code, but the variation in the results will be comparatively trivial and more certainty can always be achieved by increasing the number of samples generated.

R code (*continued*)

```
ncompare<-99999;
m<-0*c(1:ncompare);
count<-0
for (j in 1:ncompare){
      m[j]<-max(rchisq(n,1));
      }
prob<-(length(which(m>=maxgdiff))+1)/(ncompare+1);
prob
```

In this case the tail probability is found to be about 0.0035, which would certainly be regarded as significant. This significance raises the question as to why this category combination is in some way different to the remaining category combinations. These 13 ladies are having their first child, are conscientious users of the antenatal clinic, and are also making daily visits to friends by bus. One possible explanation is that these ladies are still at work. If true, then there is an unmeasured relevant fifth variable, namely employment.

Whenever there is an outlier cell, it is likely to make the associations between the variables appear more complicated than they really are. In this case, use of stepwise selection with a zero weight for the offending cell results in the selection of the much simpler model $AB/AC/BD/X$ with X denoting the outlier-exclusion dummy variable, and with the AB interaction only being included because A and B are the background factors (see Section 13.4).

The fits of the models selected, and the differences between the selected models are summarized in Table 15.8 using both the $AICc$ statistic (see Section 9.8), and the changes in G^2 with the corresponding change in degrees of freedom. The simplified model (with the outlier cell) is clearly superior to the model originally selected: it fits the data better and uses fewer parameters.

TABLE 15.8   Selected models for the data of Table 15.7

Model	$AICc$	d.f.	G^2	Change in model	d.f.	G^2
AB/AC/BC/BD/CD	161.2	11	29.29			
AB/AC/BC/BD/CD/X	142.4	10	14.90	Outlier removal	1	14.39
AB/AC/BD/X	122.0	13	19.97	Model simplification	3	5.07

The parameter estimates for the model $AB/AC/BD/X$ reveal that less use (than would be expected under independence) is made of the clinic by those mothers who have already had a child, while visits to friends by bus are made much less often than would have been predicted by the independence model. These are commonsense conclusions, but the picture would have been clouded if the outlier cell had been retained.

CHAPTER 16

PANEL DATA AND REPEATED MEASURES

Data sets of particular interest occur when every classifying variable has the same categories as every other variable. Such data may be referred to as *repeated measures* data. An example of interest to social scientists is a comparison of the social classes occupied by successive generations of the same family.

Often the data arise when a sample of individuals are interviewed at two or more time points: the individuals concerned are referred to as a *panel* and the data are referred to as *panel data*. Thus market researchers may be interested in the brand loyalty of shoppers, while political scientists will be interested in the political allegiance of voters at successive elections.

It is the case that the social classes of sons are predominantly the same as those of their fathers. Similarly, a person's preferred breakfast cereal in January of one year is likely to be the same as that a year later, and a person is likely to continue supporting the same political party. As a result, successive inquiries about class, eating habits, political allegiance, or whatever, will very often result in the same response. When the responses are cross-tabulated, the result will be a table with large numbers on the leading diagonal and relatively small numbers elsewhere. This is a situation greatly different from that of

Categorical Data Analysis by Example, First Edition. Graham J. G. Upton.
© 2017 John Wiley & Sons, Inc. Published 2017 by John Wiley & Sons, Inc.

independence and requires special models to capture the interdependence of the classifying variables. Some of these models are introduced in this chapter.

16.1 THE MOVER-STAYER MODEL

This simple model, suggested by Blumen, Kogan, and McCarthy (1955), proposes that there are two types of individuals: *movers*, whose category at the second time point is independent of their category at the first time-point, and *stayers* who stay in the same category for ever.

Since the off-diagonal cells are populated only by movers, moving independently of their previous preferences, the frequencies in these cells will follow a quasi-independence model. Studying their fitted values allows one to deduce the contributions to the on-diagonal cells from movers and, hence, by subtraction, the numbers of stayers.

Table 16.1 gives an example of a data set that perfectly fits the mover-stayer model. The numbers of stayers can be deduced from the figures given in the complete table in the following way. Consider any pair of rows (or columns) and note that, because of independence, the ratio of the observed frequencies for any pair of cells in the same column (or row) will be the same, unless one of the pair is an on-diagonal cell. For example, the ratios of the cell frequencies in rows three and four must all be equal:

$$\frac{10}{20} = \frac{20}{40} = \frac{m_3}{80} = \frac{60}{m_4},$$

where m_3 and m_4 are the numbers of movers on the third and fourth diagonal cells. To preserve the 1 to 2 ratios across the rows, the values of m_3 and m_4 must be 40 and 120, respectively. By subtraction the numbers of stayers are confirmed as $(250 - 40) = 210$, and $(240 - 120) = 120$.

The mover-stayer model often appears to fit well, but should always be treated with caution, since it may lead to a negative estimate of the number of stayers, and may lead to very different estimates of the numbers of stayers dependent on which pair of time points are compared.

TABLE 16.1   A table displaying a perfect fit to a mover-stayer model

Complete table				Movers				Stayers			
120	10	20	30	5	10	20	30	115			
10	140	40	60	10	20	40	60		120		
10	20	250	60	10	20	40	60			210	
20	40	80	240	20	40	80	120				120

TABLE 16.2   Cross-tabulation of the reported political allegiances of voters in England for the UK General Elections of 2010 and 2015

2010 vote	2015 vote				
	Con.	Lab.	Lib. Dem.	Green	UKIP
Conservative Party	4375	245	143	52	839
Labour Party	287	3068	114	131	252
Liberal Democratic Party	544	1113	957	355	355
Green Party	26	48	18	74	9
United Kingdom Independence Party	84	49	10	11	342

Source: Reproduced with permission of the British Election Study.

Example 16.1   Voters in England and the UK 2010 and 2015 General Elections

Table 16.2 summarizes data from the British Election Study concerning the reported political affiliations in England in the General Elections of 2010 and 2015.

Since the cells on the leading diagonal will be reproduced by the model, all that is required is to fit the quasi-independence model to the remaining cells. This is easily achieved by assigning zero weights to the on-diagonal cells. The required R code is:

R code

```
observed<-c(0,245,143,52,839,287,0,114,131,252,544,1113,
    0,355,355,26,48,18,0,9,84,49,10,11,0);
w<-c(0,1,1,1,1,1,0,1,1,1,1,1,0,1,1,1,1,1,0,1,1,1,1,1,0);
vote<-data.frame(expand.grid(
    e2015=c("Con","Lab","LD","Green","UKIP"),
    e2010=c("Con","Lab","LD","Green","UKIP")),
    counts=observed);
model<-glm(counts~e2015+e2010,data=vote, weights=w,
    family="poisson")
```

Unsurprisingly the model is a poor fit ($G^2 = 985.8$; d.f. $= 4 \times 4 - 5 = 11$). To see why, we compare the observed and fitted frequencies using:

R code (*continued*)

```
round(observed-model$fitted*weights,0)
```

TABLE 16.3   Residuals for the mover-stayer model fitted to the data of Table 16.2

	2015 vote				
2010 vote	Con.	Lab.	Lib. Dem.	Green	UKIP
Conservative Party	0	−282	−8	−110	401
Labour Party	22	0	10	19	−50
Liberal Democratic Party	−58	287	0	101	−330
Green Party	0	13	8	0	−20
United Kingdom Independence Party	36	−17	−9	−9	0

The results are summarized in Table 16.3. The model underestimates the number of transfers of Liberal Democrat supporters to the Conservatives and the Greens and also the transfers of Conservative supporters to UKIP. Note that the on-diagonal residuals are automatically zero.

16.2 THE LOYALTY MODEL

A simple alternative to the mover-stayer model is the loyalty model, given by

$$v_{ij} = \mu + \lambda_i^A + \lambda_j^B + \delta_D \lambda^D, \tag{16.1}$$

with $\delta_D = 1$ if $i = j$, and $\delta_D = 0$, otherwise.

This is the independence model once again, but with the on-diagonal cells treated as a special case. The anticipation is that λ^D will be positive (corresponding to on-diagonal cells having larger frequencies than would have been expected under independence).

Example 16.2   Occupational classes of two generations in the United Kingdom

Table 16.4 shows data on the occupations of males in the United Kingdom who were aged between 50 and 59 in the autumn of 2015, cross-tabulated against the occupations of their family's main wage-earner when the respondent was aged 14 (i.e., around 1975). There are major changes between the generations: the number of skilled workers has dropped considerably, and there is a marked rise in the number of professionals.

The loyalty model allows for changes in the proportions belonging to the various categories, but also takes account of the inflated on-diagonal counts where successive generations follow similar occupations. The occupation numbering is that of the survey, but the revised ordering, and the presentation

TABLE 16.4   Occupational classes of males aged 50–59 compared to the occupational classes of the main wage-earner when the respondent was aged 14

Occupational class of male respondent	Occupation of main wage earner when male respondent was 14 years old							Total
	1	2	3	5	4	6	7	
1 Managers	127	170	52	29	141	78	61	658
2 Professionals	172	440	93	56	336	136	113	1346
3 Administrators	13	43	19	8	54	16	22	175
5 Service workers	18	44	9	19	50	42	28	210
4 Skilled workers	66	101	40	47	304	138	93	789
6 Machinists	36	57	14	33	150	110	89	489
7 Elementary workers	31	37	10	17	104	75	53	327
Total	463	892	237	209	1139	595	459	3994

Source: SN 7842, Quarterly Labour Force Survey, July–September, 2015. Reproduced with permission of the UK Data Service.
The category numbers are those of the study which should be consulted for a full description of the occupational categories.

as three groups, resulted from a consideration of the residuals (given later in Table 16.5) of fitting the loyalty model.

The following R code fits the model:

R code

```
obs<-c(127, 170, 52, 29, 141, 78, etc, 75, 53);
occupy<-data.frame(expand.grid(now=factor(1:7),
    then=factor(1:7)),counts=obs);
occupy[,"loyalty"]=c(rep(c(1,rep(0,each=7)),times=6),1);
model2<-glm(counts~now+then+loyalty,data=occupy,
    family="poisson");
```

To understand the limitations of the fit of the loyalty model, we can examine the residuals using the following R command:

R code (*continued*)

```
round(observed-model2$fitted,0)
```

The residuals, which are summarized in Table 16.5, show a distinctive pattern that is frequently encountered when the categories are ordered: transitions between categories at one end of the ordered categories to categories at the

TABLE 16.5 Residuals for the loyalty model applied to the data of Table 16.4

Occupational class of male respondent	Occupation of main wage earner when male respondent was 14 years old						
	1	2	3	5	4	6	7
1 Managers	5	**47**	13	−5	−31	−15	−13
2 Professionals	**36**	**29**	17	−10	4	−45	−30
3 Administrators	−6	9	0	−1	7	−10	2
5 Service workers	−5	3	−4	−1	−7	11	3
4 Skilled workers	−10	−32	−2	10	−16	**37**	13
6 Machinists	−15	−32	−14	8	25	−7	**35**
7 Elementary workers	−4	−24	−9	0	19	**29**	−10

opposite end are comparatively rare. Managers in one generation beget managers in the next generation, but relatively few elementary workers. As the saying goes "like father, like son."

16.3 SYMMETRY

In the context of panel data, with p_{ij} denoting the probability of a transition from category i to category j, the model of symmetry states that, for all i and j,

$$p_{ij} = p_{ji}. \tag{16.2}$$

With f_{ij} denoting the observed number of transitions from category i to category j, the corresponding expected frequency under the symmetry model will be

$$e_{ij} = \frac{1}{2}(f_{ij} + f_{ji}).$$

For an $I \times I$ table, the value of a goodness-of-fit statistic would be compared with a χ^2-distribution with $I(I-1)/2$ degrees of freedom. The model might be appropriate for a system in equilibrium, but would not be useful for an evolving society.

16.4 QUASI-SYMMETRY

The symmetry requirement that $p_{ij} = p_{ji}$ implies that $v_{ij} = v_{ji}$ for all i and j, where, as previously, $v_{ij} = \ln(p_{ij})$. Writing

$$v_{ij} = \mu + \lambda_i^A + \lambda_j^B + \lambda_{ij}^{AB},$$

this in turn implies that

$$\lambda_i^A = \lambda_i^B \text{ for all } i, \text{ and } \lambda_{ij}^{AB} = \lambda_{ji}^{AB} \text{ for all } i \text{ and } j.$$

Thus there are two distinct sets of restrictions being placed on the transition probabilities by the symmetry model: one is concerned with matching the attractiveness of a category at the two time points, and the other matches the structure of changes between categories so that changes in one direction are equal (in a sense) to changes in the opposite direction. Suppose we retain the second set of constraints, but relax the first, allowing categories to change in their popularity between the two periods concerned. This defines the model of quasi-symmetry:

$$v_{ij} = \mu + \lambda_i^A + \lambda_j^B + \lambda_{ij}^{AB} \text{ with } \lambda_{ij}^{AB} = \lambda_{ji}^{AB} \text{ for all } i \text{ and } j. \tag{16.3}$$

The model, which was introduced by Caussinus (1965), has the usual center or corner constraints on the various parameters and has $(I - 1)(I - 2)/2$ degrees of freedom.

The constraints imposed by the model imply that

$$e_{ij} + e_{ji} = f_{ij} + f_{ji}, \tag{16.4}$$

where e_{ij} is the expected frequency corresponding to f_{ij}. This equality between sums of expected and observed frequencies prompted Bishop, Fienberg, and Holland (1975) to suggest constructing an $I \times I \times 2$ table in which one layer is the original $I \times I$ table, and the other is its transpose. In this way, the marginal totals of the artificial third variable C will be precisely the quantities $f_{ij} + f_{ji}$ that appear in Equation (16.4). The required equality with the sum of the expected frequencies is then achieved, because of Birch's result (Section 13.1), by fitting the model $AB/AC/BC$. Note that, because of the duplication of the original table, each sum appears twice and the goodness-of-fit value will therefore need to be halved.

Example 16.2   Occupational classes of two generations in the United Kingdom (*continued*)

The following R code fits the symmetry and quasi-symmetry models and presents the values of G^2 for each:

R code (*continued*)

```
trans<-t(matrix(obs,7,7));
expect<-0.5*(obs+trans);
symgsq<-2*sum(obs*log(obs/expect));
```

```
double<-c(obs,trans);
occ2<-data.frame(expand.grid(now=factor(rep(1:7)),
    then=factor(rep(1:7)),dummy=factor(1:2)),
    counts=double);
qs<-glm(counts~now*then+now*dummy+then*dummy,data=occ2,
    family="poisson");
qsgsq<-0.5*as.numeric(qs[10]);
symgsq;
qsgsq
```

Unsurprisingly, the symmetry model is useless ($G^2 = 321.6$, 21 d.f.). By contrast, the quasi-symmetry model ($G^2 = 21.1$, 15 d.f.) provides a very acceptable fit to the data. It is instructive to examine the estimated values of the interaction parameters for this model. Using the first category as the reference category, the estimates are shown in Table 16.6.

TABLE 16.6 The estimates for the λ_{ij}^{AB} parameters for the quasi-symmetry model applied to the data of Table 16.4. The first occupation category has been used as the reference category

	2	3	5	4	6	7
2	0.6	0.4	0.4	0.2	0.4	0.2
3		1.0	0.7	0.5	0.4	0.2
5			1.5	1.2	1.0	1.3
4				1.4	1.1	1.6
6					1.5	1.4
7						1.6

There is a clear pattern: in each row or column the on-diagonal entry is large compared to the other entries in that row or column, and, on the whole, the values get smaller as they get more distant from the main diagonal. These are typical findings when the categories of the classifying variable are ordered; they suggest that a model that takes account of the "distance" between categories may be relevant.

16.5 THE LOYALTY-DISTANCE MODEL

This model was used by Upton and Sarlvik (1981) in the context of change in political allegiance in a system governed by a single left–right political axis. The model, which can be used with any variable having ordered categories, combines the positive on-diagonal (repeated-choice) λ^D dummy parameter with one or more negative "distance-effect" dummy parameters. The latter reduce the probabilities of large changes between the categories selected at successive time points.

Example 16.2   Occupational classes of two generations in the United Kingdom (*continued*)

The occupations in Tables 16.4 and 16.5 were presented using three groups that were suggested by the pattern of signs of the residuals shown in Table 16.5. For convenience, the three groups will be labelled as U (occupations 1 and 2), V (occupations 3 and 5), and W (occupations 4, 6, and 7). We will represent a shift from Group U to group V (or vice versa) using the dummy variable E, and a shift from group V to group W (or vice versa) using the dummy variable F. A shift between groups U and W will require both dummy variables.

With δ_D defined as for the loyalty model, the loyalty-distance model is

$$v_{ij} = \mu + \lambda_i^A + \lambda_i^B + \delta_D \lambda^D + \delta_E \lambda^E + \delta_F \lambda^F, \qquad (16.5)$$

with the usual centered or cornered constraints on the main-effect parameters. The parameters λ^E and λ^F correspond to the two distance dummy variables. Labelling the categories of each dummy variable by 0 (variable absent) and 1 (variable present), their application will be as set out in Table 16.7.

TABLE 16.7   The categories of the two dummy variables used to identify the occupation transitions affected by distance effects in a three-group loyalty-distance model applied to the data of Table 16.4

Variable E							Variable F						
0	0	1	1	1	1	1	0	0	0	0	1	1	1
0	0	1	1	1	1	1	0	0	0	0	1	1	1
1	1	0	0	0	0	0	0	0	0	0	1	1	1
1	1	0	0	0	0	0	0	0	0	0	1	1	1
1	1	0	0	0	0	0	1	1	1	1	0	0	0
1	1	0	0	0	0	0	1	1	1	1	0	0	0
1	1	0	0	0	0	0	1	1	1	1	0	0	0
Transitions							Transitions						
between U and (V or W)							between (U or V) and W						

TABLE 16.8   Comparison of the fits of alternative models for the occupation class data of Table 16.4

Model	AICc	d.f.	G^2	Change in d.f.	Change in G^2
Independence	707.9	36	397.1		
Loyalty	529.6	35	204.4	1	192.7
Loyalty-distance (3 groups)	414.0	33	80.2	2	124.2
Loyalty-distance (6 groups)	400.3	29	45.4	4	34.8
Symmetry	743.7	21	321.6		
Quasi-symmetry	479.6	15	21.1	6	300.5

Some example transitions are as follows:

Within group 1, e.g., class 2 to class 1 $\qquad v_{21} = \mu + \lambda_2^A + \lambda_1^B$

Between groups 1 and 2, e.g., class 1 to class 3 $\quad v_{13} = \mu + \lambda_1^A + \lambda_3^B + \lambda^D$

Between groups 1 and 3, e.g., class 1 to class 4 $\quad v_{14} = \mu + \lambda_1^A + \lambda_4^B + \lambda^D + \lambda^E$

Between groups 2 and 3, e.g., class 7 to class 5 $\quad v_{75} = \mu + \lambda_7^A + \lambda_5^B + \lambda^E$

Table 16.8 shows the goodness of fit of the models considered in this chapter. Each model is a highly significant improvement on its predecessor. Included in the table is the model that includes separate distance parameters, one for each pair of successive occupation classes. For this model there is a tail probability of about 2.5% suggesting a model that provides a reasonable fit that is by no means perfect.

REFERENCES

Blumen, I., Kogan, M., and McCarthy, P. J. (1955) *The Industrial Mobility of Labor as a Probability Process*, Volume 6: Cornell Studies of Industrial and Labor Relations, Cornell University Press, Ithaca, NY.

Upton, G. J. G. and Sarlvik, B. (1981) A loyalty-distance model for voting change. *J. R. Stat. Soc. A*, **144**, 247–259.

APPENDIX

R CODE FOR COBWEB FUNCTION

Readers are welcome to adapt the following code as they please. The arguments of the function are: df (the data frame containing the cross-tabulation), scale (a number, usually 1, that controls the width of the cobweb lines), and outfile (the file containing the resulting cobweb diagram).

```
cobweb<-function(df,scale,outfile){
charwidth<-5.5;
nvar<-dim(df)[2]-1;
ncat<-matrix(NA,nvar,1);

Numbers of categories per variable, and required array sizes
for (i in 1:nvar){
    ncat[i,1]<-dim(table(df[i]))
}
allcat<-sum(ncat[,1]);
maxcat<-max(ncat);
x<-matrix(NA,nvar,maxcat);
y<-matrix(NA,nvar,maxcat)
z<-matrix(NA,maxcat,maxcat);
namecat<-matrix(NA,allcat,1);
ang<-matrix(NA,allcat,1);
```

```
   Lengths of variable names
varname<-names(df)[1:nvar];
lengthvar<-nchar(varname);

   Lengths of variable category names
lname<-c(1:sum(ncat));
nnames<-1;
for (i in 1:nvar){
   lname[nnames:(nnames+ncat[i]-1)]<-nchar(levels(df[,i]));
   nnames<-nnames+ncat[i];
}
   Preamble for Postscript file
heading=";%!
%%BoundingBox: 100 350 500 750
/smallfont
/Times-Roman findfont 12 scalefont def
/largeitfont
/Times-Italic findfont 16 scalefont def
smallfont setfont";

write(heading,outfile)
gap<-10.0;
rad<-150.0;
xnode<-9;
ynode<-2;
mthresh<-1;
noderad<-10;
width<-0.02;
pid<-pi/180.0;

   Range available per variable
degs<-360.0/nvar-10.0;
d1<-0.5*gap;
yc<-550;
xc<-300;
arg<-pid*360/nvar;

   Color segments
for (i in 1:nvar){
   d2<-d1+degs;
   locate<-paste( 'smallfont setfont 0.95 setgray ', xc,yc);
   circle<-paste(xc,yc,rad,d1,d2,' arc closepath fill');
   write(c(locate,circle),outfile,append=TRUE);
   d1<-d2+gap;
}
```

```
   Establish node locations
k<-0;
for (i in 1:nvar){
    if only two categories both nodes are at centre point
    if(ncat[i]==2){
       k<-k+1;
       ang[k] <-arg*(i-0.5);
       x[i,1]<- round((rad*cos(ang[k])+xc),2);
       y[i,1]<- round((rad*sin(ang[k])+yc),2);
       x[i,2]<-x[i,1];
y[i,2]<-y[i,1];          k<-k+1;
       ang[k]<-arg*(i-0.5);
    }

    With more than two categories
    if(ncat[i]>2){
       start<-0.5*gap+(i-1)*(degs+gap);
       degbit<-degs/ncat[i];
       halfdeg<-0.5*degbit;
       for (j in 1:ncat[i]){
          k<-k+1;
          ang[k]<-(start+halfdeg+(j-1)*degbit)*pid;
          x[i,j]<-round(xc+rad*cos(ang[k]),2);
          y[i,j]<-round(yc+rad*sin(ang[k]),2);
}}}

udf<-matrix(unlist(df),dim(df)); print(udf); last<-nvar+1;

    Squared standardized residuals
for(i in 1:(nvar-1)){
    for(j in (i+1):nvar){
       z<-chisq.test(xtabs(as.numeric(udf[,last])
       ~udf[,i]+udf[,j],df))$stdres
       zsq<-z*z
       for(k in 1:ncat[i]){
          for (l in 1:ncat[j]){
             w<-0;
             if(zsq[k,l]>4){
                w<-abs(trunc(z[k,l]))*scale;
                if(z[k,l]<0){
                   write('0.85 setgray ,outfile,append=TRUE)
                }
                if(z[k,l]>0){
                   write('0 setgray newpath ',outfile,
                   append=TRUE)
```

```
            }
            write(paste(w,' setlinewidth ',x[i,k],y[i,k],
                ' moveto ',x[j,l],y[j,l],' lineto stroke ',
                ' newpath '),outfile,append=TRUE)
}}}}}
k<-0;
for (i in 1:nvar){
    if(ncat[i,1]==2){
        k<-k+1;
        namecat[k]<-row.names(table(df[,i]))[1];
        k<-k+1;
        namecat[k]<-row.names(table(df[,i]))[1];
    }
    if(ncat[i,1]>2){
        for (j in 1:ncat[i,1]){
        k<-k+1;
        namecat[k]<-row.names(table(df[,i]))[j];
}}}

    write node names
k<-0;
for(i in 1:nvar){
    for(j in 1:ncat[i]){
        k<-k+1;
        write(paste('1 setlinewidth 1 setgray ',
        x[i,j],y[i,j],noderad,' 0 360 arc fill 0 setgray '),
            outfile,append=TRUE);
        xx<-x[i,j];
        if(xx<xc){
            xx<-xx-lname[k]*charwidth;
        };
    yy<-y[i,j]+ynode;
    write(paste(xx,yy,' moveto (',namecat[k],') show ',
        ' newpath '), outfile,append=TRUE) };
write(paste('largeitfont setfont '),outfile,append=TRUE);

    Write variable names
for (i in 1:nvar){
    ang<-arg*(i-0.5);
    xx<-round((1.15*rad*cos(ang)+xc),2);
    yy<-round((1.15*rad*sin(ang)+yc),2);
    if(xx>=xc){
        xx<-xx-lengthvar[i]*5.5
    };
    if((ang>0.5*pi)&(ang<1.5*pi)){
```

```
      xx<-xx-55.0
   }
   script=paste(xx,' ',yy,' moveto (',varname[i],') show ',
   ' newpath');
   write(script,outfile,append=TRUE);
}
write(paste('smallfont setfont '),outfile,append=TRUE);
}

close='showpage'; write(close,outfile,append=TRUE);
}
```

INDEX

AUTHOR INDEX

INDEX OF EXAMPLES